RED ROCKS COMMUNITY COLLEGE

D0583391

QL
737
P93
A4

23107 - 2

Altmann, Stuart A.

Baboon ecology

Date Due

AUG 06			
DEC 06 '95			

COMMUNITY COLLEGE OF DENVER
RED ROCKS CAMPUS

BRO
DART Printed In U.S.A.

BABOON ECOLOGY
AFRICAN FIELD RESEARCH

BABOON ECOLOGY

AFRICAN FIELD RESEARCH

STUART A. ALTMANN AND JEANNE ALTMANN

The University of Chicago Press, Chicago and London

COMMUNITY COLLEGE OF DENVER
RED ROCKS CAMPUS

Distributed simultaneously in the United States, its dependencies, the Philippine Islands, the British Commonwealth, including Canada, by THE UNIVERSITY OF CHICAGO PRESS, Chicago and London

Library of Congress Catalog Card Number: 72-116763
International Standard Book Number: 0-226-01601-3 (clothbound)

The University of Chicago Press, Chicago 60637
The University of Chicago Press, Ltd., London
S. Karger AG, Basel

23107-2
QL
737
P93
A4

#106267

© 1970 by Stuart A. Altmann. All rights reserved
Published 1970
Second Impression 1973

Printed in the United States of America

COMMUNITY COLLEGE OF DENVER
RED ROCKS CAMPUS

CONTENTS

ACKNOWLEDGEMENTS

Our baboon field research and subsequent data analysis were financed by the National Science Foundation, through a series of research grants (GB 683, GB 3903, GB 2879, and GB 4415) and by U.S. Public Health Service Research Grants MH-07336-01 and FR-00165 from the National Institutes of Health. We are grateful to the Foundation and the Institutes for their generous support.

We wish to express our appreciation to the many people who contributed to our study. In field studies, assistance often comes in many and unusual forms. We are deeply grateful to the Robert Coxes for many kindnesses that contributed to the success of our study. Our field secretaries, Lindy Edwards and Dian Annear, did a wonderful job under circumstances that were often trying. The cooperation of the Game Wardens of Amboseli, first Jan Allen, then Moses S. Loilonya, made our study there possible. Diane Story, Lillian Champion, Vivian Galligan, and Kathy Russell, our statistical clerks, coped diligently with our voluminous data.

During our first month in East Africa, Irven DeVore accompanied us on several trips through the Nairobi National Park and thereby speeded our familiarization with baboon life. Thomas Struhsaker, whose research on vervet monkeys was carried out in Amboseli during the course of our study, contributed much, through discussion, joint observations, and constant good spirits.

Plant identifications were made by J. B. Gillett, Botanist-in-Charge, East African Herbarium. We are very grateful for this valuable information.

Our colleague, Stephen S. Wagner, provided invaluable assistance by developing some of our techniques for quantitative analysis of field data. In addition, he carefully read the entire manuscript and did much to clarify the presentation.

The art work for this volume was done by our Primate Center's staff artist, Mrs. Patricia Foster. Photographic prints were made by Frank Kiernan.

Finally, we are grateful to our son, Michael, without whose patience and cooperation our field work would have been much more difficult.

I. INTRODUCTION

Just north of Mount Kilimanjaro, in East Africa, there is a vast, semi-arid plain. The Masai-Amboseli Game Reserve occupies approximately 1,250 mi^2 of this area. Rain and meltwater from snow on Kilimanjaro percolate through the igneous rocks on the slopes, disappear from the surface, and flow underground. Here and there on the Amboseli plain, where surface depressions dip below the water table, natural springs have formed. The resulting waterholes and swamps, with their surrounding vegetation, make possible a remarkable concentration of African wildlife.

Amboseli was the major site of a field study on baboons which we carried out between June, 1963, and August, 1964. Although our main interests were in baboon behavior and social organization, we gathered much ecological data as well, both because of the unusual opportunities to do so and because any adaptive or evolutionary explanation of the behavior of an animal necessarily involves an understanding of the animal's relations with its environment.

The present monograph covers the major ecological observations that were made during our study. In subsequent reports we shall describe the behavior, social organization and intergroup relations of the baboons. That is, we shall be concerned here with the ways in which a baboon social group, as a unit, meets the challenges of life on the savannahs of East Africa, and in subsequent reports, with the internal relations and organization of the group that enable the animals to cope with these external problems.

This is a success story. Baboons, found throughout much of subsaharan Africa, are among the most widespread, abundant and adaptable of primates. They live in a variety of habitats, ranging from subdesert steppe, through savannahs with various proportions of woodland, to moist, evergreen forests. This monograph describes how baboons make a living. It is based primarily on the animals in the Amboseli Reserve, with comparisons to baboons elsewhere in Africa.

After a look at study areas and research methods in chapter II, and baboon populations in chapter III, we trace the activities of one group of Amboseli baboons, beginning with sleep, awakening, and descent from the trees (chapt. IV). The movements of the group on their daily round are described in chapter V, followed by a chapter on food and water. Relations with predators are described in chapter VII, which precedes a chapter on the baboons' relations with other species of animals. Finally, in chapter IX we offer a number of speculations about baboon ecology and adaptations, and point out some of the implications of these ideas for further research.

1. Baboon Species

The taxonomic relations among the various populations of African baboons is still rather poorly known, but research on these animals is expanding rapidly. As a result, their classification is now in a state of rapid flux. Even the term 'baboon' is used in various ways. It may be used, as JOLLY [1967] proposes, for any predominantly terrestrial cercopithecids (old-world monkeys), i.e. for animals in a similar adaptive zone but only broadly related taxonomically. As such, the term would include geladas, hamadryas, the 'typical' baboons, and patas monkeys. Or the term may be confined to the genus *Papio*.

But what animals should be included in *Papio*? Just mandrills and their congeners, says HOPWOOD [1947], thereby excluding both hamadryas and 'typical' baboons. No, hamadryas and 'typical' baboons but not drills or mandrills, says HILL [1967]. All of the above and geladas, too, says BUETTNER-JANUSCH [1966]—and throw in the Celebes 'macaques' and the Celebes black 'ape', says ROTH [1965].

The problem continues at the species level, particularly in the classification of the 'typical' or common baboons of the African savannahs. They are all one species, from the Cape to the Sahara, say DEVORE and WASHBURN [1963], ROTH [1965] and BUETTNER-JANUSCH [1966]. No, the yellow and the anubis baboons are each extremely homogeneous populations, with a broad geographic gap between them, say MAPLES and McKERN [1967].

There seems to be no disagreement that the baboons that we studied in the Amboseli area are to be classified as 'yellow' or 'cynocephalus' baboons (Gr. *kynos* = dog, *kephalikos* = head). In comparing our observations on these animals with observations made elsewhere, either by us or by others, we shall use the classification proposed by

JOLLY [1966 and in HILL, 1967], which divides the genus *Papio* into five species as follows:

(1) *P. hamadryas* Linn. 1758: sacred, hamadryas or mantled baboon,
(2) *P. papio* Desmarest 1820: Guinea baboon,
(3) *P. anubis:* anubis or olive baboon,
(4) *P. cynocephalus* Linn. 1766: yellow baboon,
(5) *P. ursinus* Kerr 1792: chacma baboon.

Distinguishing external characteristics and geographical distribution of these species are given by JOLLY [1966]. Our use of this classification is one of convenience, and does not imply any taxonomic judgement on our part[1].

2. Other Baboon Field Studies

In recent years, each of these species has been studied in its natural habitat. Comparison of yellow baboons with the other baboon species will be made in our text. The chacma baboon was studied in Kruger National Park, South Africa, by BOLWIG during 1953, in various parts of South-West Africa and S. Rhodesia by HALL, and in South Africa by HALL and WINGFIELD in a series of studies between 1958 and 1961 [HALL, 1960, 1962a, 1962b, 1963; WINGFIELD, 1963]. Further studies on chacma baboons in the northern Transvaal are now being carried out by LUKAS P. STOLTZ and GRAHAM SAAYMAN [1969]. Brief observations on chacma baboons in South-West Africa were made during 1968 by W. J. HAMILTON [personal communication].

The olive or anubis baboon was studied by MAXIM and BUETTNER-JANUSCH[2] during 1960 [MAXIM and BUETTNER-JANUSCH, 1963], by

---

[1] African names for baboons may be of use to field workers. We do not know whether any African languages distinguish the various species of baboons. Some African names for baboons are as follows: *hoku* in Kikami, *nyabu* in Kisagara, *nyani* in Kiswahili and Nyanja, *mhuma* in Chicogo, *pooma* in Kinyaturu, *kuku* in Kisukuma [LOVERIDGE, 1923], *dayer* in Somali [DRAKE-BROCKMAN, 1910], *kolwe wa mpili* in Bemba, *pombwe* in Lozi, *mpombo* in Kaonde, *sokwe* or *pombo* in Tonga [ANSELL, 1960], *otolal* in Masai.

[2] MAXIM and BUETTNER-JANUSCH indicate that their animals were *P. doguera* (= *P. anubis*). However, according to MAPLES and MCKERN [1967], the baboons in the area where MAXIM and BUETTNER-JANUSCH worked are *P. cynocephalus*.

HALL in Uganda during 1963 [HALL, 1965a and b], by ROWELL in
Queen Elizabeth National Park, Uganda, between 1963 and 1965
[ROWELL, 1964, 1966], by CROOK and ALDRICH-BLAKE [1968] near
Debra Libanos, Ethiopia, in 1965, by ALDRICH-BLAKE [personal com-
munication] in the Awash Valley of Ethiopia during 1968, by WARSHALL
in Nairobi Park, Kenya during 1964 [WARSHALL, personal communi-
cation], by DEVORE and WASHBURN in Nairobi National Park, Kenya
during 1959 and 1963 [DEVORE, 1962, 1963; DEVORE and WASHBURN,
1960, 1963; WASHBURN and DEVORE, 1961] and by WASHBURN in the
Wankie Game Reserve, Rhodesia [WASHBURN and DEVORE, 1961].
Baboons of the Wankie Reserve and in Matopos National Park, S. Rhodesia, were observed by TUTTLE and his students in a brief study during
1965 [CARTMILL and TUTTLE, 1966; MORGAN and TUTTLE, 1966]. We
made some observations on this species in 1963, during our initial
reconnaissance, as did STRUHSAKER [personal communication] during
1964 in the course of his work on vervet monkeys.

The study by DEVORE and WASHBURN included some observations
on yellow baboons in the Amboseli Reserve and Tsavo National Park,
Kenya; WARSHALL [personal communication] worked on baboon
ecology in Amboseli during 1964. A brief survey of baboon groups in
parts of Tsavo was made by STRUHSAKER [personal communication] in
1964. Our study was apparently the first intensive field work on this
species.

The hamadryas baboon was studied in Ethiopia by KUMMER and
KURT during 1960–61 [KUMMER and KURT, 1963; KUMMER, 1967, 1968].
At the time of this writing, KUMMER is once again in Ethiopia, con-
tinuing this research.

So far as we know, the recent reports by BERT, AYATS, MARTINO
and COLLOMB [1967a and b] represent the only contemporary field work
on the Guinea baboon[3].

The literature on adaptations of baboons has recently been
brought together in reviews by HALL [1966], HALL and DEVORE [1965],
DEVORE and HALL [1965], and JOLLY [1967], covering baboon behavior,
ecology, morphological adaptations, distribution and evolutionary history. Over the years, baboons have also been the subject of many
anecdotes, tales, and unsystematic observations. The most informative

---

[3] According to TAPPEN [1960], the specific distinction between Guinea baboons
and olive baboons is highly questionable.

accounts are by STEVENSON-HAMILTON [1947] and FITZSIMONS [1919]. The most dramatic is by MARAIS [1947]. The history of man's relations with baboons (e.g. as house pets in ancient Egypt) may be found in MORRIS and MORRIS [1966] and in MACDONALD [1965].

II. STUDY SITES AND METHODS

1. Reconnaissance Safari

We arrived in East Africa early in June, 1963. During the first six weeks after our arrival, while we were making the necessary arrangements for our extended field study, we made occasional observations on baboons in Nairobi National Park, Kenya. Then, during the month that followed, we carried out a survey of baboon populations and habitat in southern Kenya and northern Tanganyika (now Tanzania). This involved a trip of over 2,000 mi. by Landrover, most of it through savannah country (fig. 1).

Each area that was visited during this reconnaissance trip was compared with Nairobi National Park, which seemed beforehand to be the most favorable site for our study because of background data that resulted from DeVore's previous field work on baboons in the area. Our comparisons of potential study areas were based on the density, approachability, and visibility of baboons and associated animals, as well as access to gasoline, food, and other supplies, availability of outside communication and mail service, availability of a local mechanic, driving conditions both on and off established roads or tracks, availability of water for drinking and bathing, and the cooperativeness of local game wardens and other personnel.

We summarize below our observations on the wildlife reserves that we visited during our trip, pointing out some of the major advantages and disadvantages of each for naturalistic studies of baboon behavior. We did not find baboons in any of the areas outside of wildlife reserves except for a few in the vicinity of Olnaigon Swamp, Kenya. In part, this was because the wildlife in general has been killed off in these areas. Beyond that, however, we did not tarry in such areas because we knew how impractical it would be to set up any long-term studies in them. Thus, we may have missed small, inconspicuous populations.

Nairobi National Park (various dates, June 21 to July 21). This park, on the outskirts of the city of Nairobi, is an excellent place in which to see a wide variety of African wildlife. The baboons of this park were first studied by Irven DeVore during 1959. Fortunately, DeVore was back in the area at the time that we began our study, carrying out a recensus of the Nairobi Park baboons [DeVore, 1965]. With his guidance, we quickly became familiar with many of the habits of these animals. Thus, Nairobi Park would have offered the great advantage of a previously studied baboon population as well as proximity to sources of supplies. But it was populated with baboons that were so accustomed to being fed by tourists that they responded continually to the presence of any human being, a problem that had become more severe since 1959, according to DeVore. This concentrated supplemental food produces

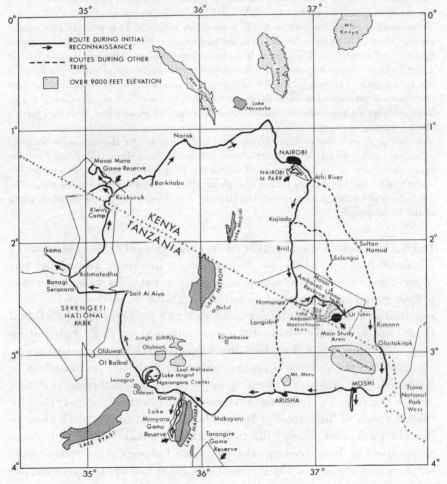

Fig. 1. Route of initial reconnaissance (solid line) and of subsequent trips (dashed lines).

conspicuous changes in the behavior of the animals and doubtless in their habitat utilization as well.

Masai-Amboseli Game Reserve, Kenya (July 24 to August 7). In contrast with the irregular terrain of Nairobi Park, much of the southern portion of the Amboseli Reserve is quite flat. The flatness is relieved by a few well-demarcated hills, made up of igneous extrusions that are probably related to the vulcanism of Mount Kilimanjaro.

The habitat within the study area is semi-arid acacia savannah (fig. 2). It has been described by STRUHSAKER [1967] as follows:

'The Masai-Amboseli Game Reserve lies in the plains immediately to the north of Mount Kilimanjaro and is located 2° 40′ S, 37° 10′ E, with an elevation of about 3,700 feet. Temperatures range from 48 to 90 °F, sometimes within a 24-h period. The annual precipitation of 10 to 20 inches is restricted to 2 periods: November through December and March through May. The habitat at Amboseli is typified as semi-arid savanna, having a small number of plant species. Permanent water holes and swamps within the Reserve are fed by springs arising from the underground drainage of Mount Kilimanjaro. Near these permanent sources of water there are relatively dense groves of fever trees *(Acacia xanthophloea)* with an understory dominated by *Azima tetracantha* and *Salvadora persica* shrubs. Living in these groves were greater concentrations of vervet monkeys than I observed anywhere else in East Africa. Other primates in the Reserve were baboons *(Papio cynocephalus),* bush babies *(Galago senegalensis),* and Masai tribesmen. Furthermore, there were at least 18 species of ungulates, 15 species of carnivores, and a wide variety of birds. All animals within the Reserve were protected against hunting. The only source of human disturbance to the Reserve was overgrazing by the domestic stock of the Masai people.'

During our reconnaissance, this reserve was the only place, other than Olnaigon Swamp, where we found yellow baboons, and all of our observations confirm the sharp demarcation between yellow baboons *(Papio cynocephalus)* and olive baboons *(Papio anubis)* that has been reported and mapped by MAPLES and McKERN [1967]; that is, we found these two forms only within the areas indicated on their distribution map, and in no case did we find any baboons between these two areas.

Within the vicinity of permanent water, baboons are abundant in Amboseli. We censused several groups in Amboseli during this first visit to the area. The baboons in Amboseli are relatively undisturbed by the approach of humans. Far fewer tourists come to Amboseli than to Nairobi Park, and, except for two or three animals that had become accustomed to hand-feeding, the Amboseli baboons were conspicuous for their lack of concern about reasonably quiet and unobtrusive human observers.

Fig. 2. Habitat of the Masai-Amboseli Game Reserve, Kenya. (a) Fever trees *(Acacia xanthophloea)*, (b) Umbrella tree *(A. tortilis)*.

Gasoline and vehicle repairs can be obtained within the Amboseli Reserve at Ol Tukai. Supplies can be brought in from Arusha or Nairobi, and there is a small staples store in the Reserve. A commercial tent camp is available in Ol Tukai but is prohibitively expensive for a long-term study. In addition, a lodge for tourists was built at Ol Tukai during our stay in Amboseli.

Another advantage of Amboseli to us at the time was that THOMAS STRUHSAKER, then a graduate student in zoology, was carrying out an extensive study of behavior and ecology in the Amboseli vervet monkeys, *Cercopithecus aethiops*.

Lake Manyara National Park, Tanganyika (August 9–10). Baboon groups were seen only in the ground-water forest, at the north end of the park (fig. 3); none were seen in the acacia woodland or scrub area in the central and southern portions. Vervet monkeys and blue monkeys *(Cercopithecus mitis)* were also seen in this Park.

During 1962, 40–50 baboons were live-trapped in the Park; 700 others were killed by means of firearms. The justification given for this

Fig. 3. Ground-water forest in Lake Manyara National Park, Tanzania.

in reports of a previous park warden [MORGAN-DAVIES, 1961, 1962] to the Director of the Tanganyika National Parks is that the abundance of baboons in the ground-water forest of the Park is detrimental to the bird population there. The basic evidence offered in support of this theory is that birds are more abundant in the open grassland portions of the Park, where baboons are seldom found. We were not able to find any indication of an analysis of the ecology or population dynamics either of the baboons or of any species of birds living in Lake Manyara Park.

Of course, that action made this Park completely unusable for our naturalistic study.

Ngorongoro Crater, Tanganyika (August 11–13). Only one group of baboons was found within this old volcanic crater. It was on the southwest edge of the Lerai Forest, on the crater floor. In addition, we heard reports of baboons living on the forested slopes of the crater. In any case, it was made clear to us that we would not be permitted to camp for any extended period within the crater, and thus that any extensive studies of baboons in the area would be impossible.

Serengeti National Park, Tanganyika (August 13–19). Large areas of Serengeti consist of immense tracts of flat, treeless grassland. We did not see baboons in any of this area. About 16 mi. east of Seronera there are a number of large, rocky outcroppings ('kopjes'). Baboons were found on a large kopje, next to the road (fig. 4). The kopjes are surrounded by treeless, grassy plains. The baboons seemed quite at home, scampering about on the rocks.

Streams and river forest are abundant in the vicinity of Seronera. We found a number of baboon groups there, including the smallest baboon group that we saw anywhere in East Africa, consisting of only 10 animals. We found baboons more difficult to locate and to approach in the Serengeti than in Amboseli or Nairobi. However, the relations between the local baboons and the migratory ungulates [BARTLETT and BARTLETT, 1961] are of particular interest and warrant a thorough study. None of the scientific staff was at the Michael Grzimek Memorial Laboratory, at Banagi, Serengeti, when we visited there, and those camp sites that the warden permitted us to use would have been unsuitable for long-term camping because of the lack of bathing facilities and the proximity to campers.

Masai-Mara Game Reserve, Kenya (August 19–23). The Mara Reserve lies north of the Serengeti, just over the Kenya border. Many of the ungulates found on the Serengeti migrate through this area

a

b

Fig.4. A rocky outcropping in Serengeti National Park, Tanzania, inhabited by olive baboons.

[GRZIMEK and GRZIMEK, 1960]. Despite extensive driving through this area, we found only one group of baboons, about 18 mi. along the road from the Keekorok Lodge to Narok. THOMAS STRUHSAKER, who spent 7 days there during the month of July, 1964, encountered baboons at 14 locations in the Reserve, with some groups probably being encountered several times. The most frequent sitings were along the Talek River, 5–10 mi. before it joins the Mara R., and near the confluences of the Mara, Sigera and Olchorro Loldabaith Rivers.

On the basis of what we saw during this initial reconnaissance, we decided that the Amboseli Reserve, in southern Kenya, was the unequivocal choice as the study site for our detailed observations. Nowhere else that we visited was there such an abundance of readily observed and relatively undisturbed baboons.

All observations from September, 1963, until the termination of the study on August 4, 1964, were made in Amboseli. After October 1, 1963, almost all observations were made on one group of baboons, which we designate in this work as the *Main Group*. As a result of several personal mishaps, observations were made only one day between December 4, 1963 and February 8, 1964, only the junior author was in the field March 23 to April 2, 1964, and only the senior author was in the field from April 21, 1964 until the termination of the study (with the exception of 2 days). During much of the rest of the time, both observers were in the field together, and many of the observations that were made by one observer were verified by the other.

In total, 1,469.2 hours were spent in field observations, including 1,006.1 h with the Main Group, of which 302.0 h were spent in joint observations, by both observers. Of this time with the Main Group, 887.1 h were spent on close observations; the remainder, 119.0 h, were spent in more casual observations on the group, including time spent on photography, mapping, making sound recordings, talking with visitors, and so forth. The sampling of data was not random through the day, particularly because we knew that social interactions, which were one of the major subjects of our expedition, were usually far more frequent late in the afternoon, before the baboons ascended their sleeping trees, and again in the morning, shortly after their descent.

Because we shall often be concerned with the rates at which various behavioral and ecological events occur, a detailed breakdown of field time by hour of the day and season is presented in table I.

Table I. Field time.

Beginning of hour	Jan-Feb		Mar-Apr		May-Oct		Nov-Dec		Annual	
	B	C	B	C	B	C	B	C	B	C
0600	0	0	0	51	0	3	0	7	0	61
0700	30	310	18	321	0	318	0	419	48	1368
0800	100	592	195	1137	151	2277	0	856	446	4862
0900	43	747	214	1571	551	3526	20	940	828	6784
1000	21	755	192	1579	677	3928	73	862	963	7124
1100	19	529	211	1444	702	3649	163	627	1095	6249
1200	30	248	81	1058	559	2525	37	581	707	4412
1300	0	37	75	803	433	1665	70	410	578	2915
1400	0	60	49	721	342	1622	65	305	456	2708
1500	30	120	97	665	342	1830	35	313	504	2928
1600	60	216	22	802	449	2588	57	428	588	4034
1700	60	554	52	1384	470	3420	0	693	582	6051
1800	30	416	52	1112	264	1869	0	313	346	3710
1900	0	0	0	20	0	0	0	0	0	20
Total, B and C time	423	4584	1258	12668	4940	29220	520	6754	7141	53226
A time	1664		4358		19148		2374		27571	

The table gives the number of minutes of observation during each season and each hour of the day. Observation minutes are classified as *A*, *B*, or *C* time. *A* time is general field time (including reconnaissance time and time spent observing groups other than the Main Group). *B* time is casual observation time on the Main Group (including times when visitors were with us, time spent on cinematography, mapping, setting up sound recorders or other equipment, and any other periods of casual watching during which we took few notes and made no systematic observations). *C* time is time spent making close observations on the Main Group.

2. Methods of Study

We tried to maintain a neutral relationship with the baboons, at no time attempting any intervention in anything they did or anything that befell them. We tried not to move our field vehicle directly toward the group, and if we could anticipate their line of progression, we tried never to stop in their pathway. We never fed anything to any of them. Gradually the baboons of the Main Group—and to a lesser extent, baboons of other, nearby groups—came to recognize our vehicle and us. Under those conditions, prolonged observations at close range were possible. We routinely worked at a distance of about 50 feet or less from the nearest member of the group.

Most of our observations were made from a long-wheelbase Landrover stationwagon that was specially modified for our field work. A major modification consisted of an observation platform built onto the roof of the vehicle. Ready access to this platform was provided by a hatch in the roof.

A standard double-bed mattress was fitted into the back of the vehicle. By suspending this mattress, with plywood and an angle-iron frame, level with the top of the middle seat, we had ready access to baggage stored underneath. When we set up a tent camp in Amboseli, the bed was removed and the back of the Landrover was converted into a playroom for our infant son.

An aerial photograph of the Amboseli study area (V13B/RAF/341, frame 158, 26 JAN 63) was obtained from the Survey of Kenya Office. A tracing was made from this photograph on which we drew the major features of the terrain; at a scale of 1:15,000 it is possible to distinguish individual acacia trees. The resulting map was divided into sections that would fit conveniently into a standard 8½ × 11 inch, loose-leaf notebook, and multiple prints were made of each section.

In the field, these map prints were used to plot positions and movements of groups. After a few weeks of practice it was usually possible to plot any point within about 25 ft. In the case of our Main Group, we made a continuous plot of the approximate course of the center of mass of the group, but drawn so that the plotted line remained within the confines of the group.

All other ecological and behavioral data were recorded in field notebooks. Data were recorded at the time of observation, with no subsequent 'amplification'. Time was recorded to the nearest minute, or to the nearest $1/10$ minute in the case of some types of behavior sampling. Many of the data were dictated into a portable recorder and subsequently typed. This technique has several advantages over handwriting notes. Without the necessity for looking repeatedly into a notebook, the observer can keep animals under continuous observation, something which is of particular importance in taking censuses and when gathering detailed data on rapidly occurring, complex processes. We gathered about five times as many data when our field recorders were operating as we did when they were not. On the days when at least one recorder was operational, we dictated about 9,000 words on the average, with a maximum of about 14,000 words. (Unfortunately, the recorders that we used, Minifon wire recorders, frequently broke down and we could not get satisfactory repairs from the factory.)

Data in these voluminous field notes are located by means of a permuted index, which was produced after our return from the field, in the following manner. As each paragraph of the field notes was read, one of us (S. A.), using an office recorder, dictated all index entries that were relevant to that paragraph, along with the identification number of the paragraph (page number[4] and time of day). This information was punched on IBM cards, then permuted, alphabetized and printed by a computer. The permuting results in an index entry under each word in the index entries. For example '*adult male*' is alphabetized not only under *a* but also under *m (male adult)*.

Special techniques that were used for particular kinds of data are described where appropriate in the chapters that follow.

In general, our goal was to understand how the animals cope with the problems that they face in their natural habitat; toward this goal, we tried to obtain records that would be adequate, in terms of accuracy of observation, quality of description, and quantity of data.

---

[4] Subsequent work with the index indicates that a date code would be preferable to page number.

III. POPULATIONS

Like virtually all other primates, the baboons of Amboseli live in discrete social groups. These groups are semi-closed: except through birth or death, individuals seldom enter or leave them. As a result of this stability, the social group is a basic population unit as well as a practical unit for census work.

After describing our techniques of censusing free-living baboon groups we shall present our data on the size and composition of the groups that we censused in Amboseli and elsewhere. These data will be compared with those in the literature.

1. Census Techniques

Our technique for taking censuses of baboon groups was as follows. We positioned ourselves in such a way that we could watch the baboons pass an arbitrarily selected 'counting point' that was clear of obstructing foliage. This was best done in the morning as the baboons were moving in file toward the foraging areas. Binoculars, often held in place by means of a small tripod, were kept aimed at the counting point, and one person kept continuous watch on the counting point while the other either took dictation (at times when our magnetic wire recorders were not working) or verified the counts and the age-sex determinations. A rear oblique view of the animals as they passed was best, since it gave a view of the ischial callosities (which are separate only in females) and the sexual skin.

As each individual passed the counting point, the observer indicated its age class, sex, name or conspicuous identification marks, and, when possible, the condition of the females' sexual skin. If an infant was riding on its mother's back or belly, this was indicated. The age classes that were used are given in table II.

At the end of each census, we indicated the overall quality of age-sex determinations, the estimated percent of observations that were verified by the second observer, and an 'error factor', which gives estimates of, first, the number of animals that might have been missed (due to obstructing foliage and so forth), and second, the number of animals that might have been counted twice (usually the result of young animals running back and forth across the counting point). For example, 'error factor (+2, -1)' means that two animals may have been missed and one may have been counted twice. As a further check on census accuracy, simple counts of individuals were sometimes taken on groups. Simple counts were also made when observational conditions did not permit a more detailed census.

In Amboseli it was important not to census animals too early in the morning, along the major routes leading from concentrations of sleeping trees, for at that time of day two or more groups often followed one after the other in a long procession, sometimes with the rear member of one group closer to the front member of the following group than to any member of his own group. At such times, unless one is censusing recognizable individuals, it is important to wait until the groups have diverged and begun to move independently. On the other hand, censuses are usually very difficult later in the day, when the members of each group are spread out during foraging. The ideal time, then, is after the groups have diverged, but while the animals are still moving in file.

Table II. Age-sex classes and their characteristics, yellow and anubis baboons

Class	Age in years	Physical characteristics
Infant-1	0– ½	Hair completely or partially black (natal coat). Skin pink or red from skin vascularity
Infant-2	½–1	Hair brown to cream-colored, often lighter (in Amboseli) than that of adults. Skin pigmented black, as in adults
Juvenile-1	1–2	Not sharply demarcated from previous class. Light hair retained in yellow baboons
Juvenile-2	2–4	Not sharply demarcated from previous class. Hair often darker, as in adult
Adult female	over 4	Sexually mature; sexual skin swells periodically. Nipples button-like when nulliparous, elongated in more mature, multiparous females
Subadult male	4–6	Development of secondary sexual characters: mantle (in anubis), long canine teeth, large size, greater musculature than females. Scrotum (testes) larger than in juvenile-2
Adult male	over 6	Secondary sexual characteristics fully developed

For very accurate census work, it is essential that repeated censuses be taken on each group, and that each group have several recognizable 'marker' individuals. It is not good practice to identify or distinguish groups solely on the basis of census results, for two groups would then be considered as one if they had the same composition (or at least, the same census results), and one group would be considered as several if it yielded varying census results, because of either census inaccuracies or natural changes in composition with time.

2. Census Accuracy

In the years since C. R. CARPENTER's pioneering survey of howler monkey populations on Barro Colorado Island [CARPENTER, 1934] numerous censuses have been taken on primate groups in their natural habitats. In none of these censuses, however, has there been any objective measure of the amount of sampling error or observer reliability that is involved. Such a measure can be obtained only by matching the results of censuses against the true composition of groups, as determined by some independent and more accurate method[5].

For our main study group in Amboseli the actual composition was worked out on the basis not only of repeated counts and censuses but independently on the basis of individuals that we could recognize. If some individual was not observed during a census, during a scheduled time sample on social behavior, or during other routine observations, a search for that individual was initiated in order to discover if it was still alive and with the group. Sixty-eight complete censuses were taken on the Main Group. That is to say, there were 68 censuses in which we felt that every individual in the group had been seen at the time of the census (not just thereafter, upon tallying the results). Data from numerous incomplete censuses were not used. The number of adult males and adult females at the time of each of these 'complete' censuses is known; only for the first few censuses might there be any doubt about the actual numbers. The results of each census have been compared with the known composition of the group at the time of the census. The results are shown in figure 5, which reveals the following. First, there was a clearcut tendency for the censuses to become more accurate: for both adult males and adult females there was a several-fold decrease in the

[5] SOUTHWICK and SIDDIQI [1966] have indicated the extent to which group counts vary in repeated censuses of rhesus in one area. The difference between observer agreement and observer reliability is the difference between consistency and accuracy.

Dates of change	Males							Total males	Females						Total females	Sex unknown		Total	Number of days at this composition	Ratio of adult males to adult females
	I₁	I₂	J₁	J₂	Sub	Ad	?		I₁	I₂	J₁	J₂	Ad	?		I₁	Other			
29.7.63	1	5	3	3		7		19	1		2		18		21			40	41	0.388
7.9.63	2	5	3	3		7		20	1		2		18		21			41	5	0.388
12.9.63	3	5	3	3		7		21	1		2		18		21			42	22	0.388
4.10.63	4	5	3	3		7		22	1		2		18		21			43	1	0.388
5.10.63	3	5	3	3		7		21	1		2		18		21			42	18	0.388
23.10.63	2	5	3	3		7		20	1		2		18		21			41	26	0.388
18.11.63	2	5	3	3		9		22			3		18		21			43	44	0.500
1.1.64		5	7	3		9		22			3		18		21			43	11	0.500
12.1.64	2	2	7	3		9		22			3		18		21	1		44	35	0.500
16.2.64	2	2	7	3		10		22			3		18		21	1		45	20	0.555
7.3.64	2	2	7	3		9		23			3		18		21	1		44	5	0.500
12.3.64	2	2	7	3		9		22			3		18		21	1		43	6	0.500
18.3.64	2	2	7	3		8		22			3		18		21			42	32	0.445
19.4.64	2	2	7	3		8		21			3		17		20			41	4	0.470
23.4.64	2	2	7	2		7		21			3		17		20			39	22	0.411
15.5.64	1	1	7	2		7		19			3		17		20			38	10	0.411
25.5.64	2	1	7	2		7		18			3		17		20			40	11	0.411
5.6.64	2	1	6	2		6		19	1		3		17		21			40	18	0.352
23.6.64	2	1	6	2		6		19	1		3		17		21			39	10	0.352
3.7.64	3	1	6	2		5		18	1		3		17		21			38	3	0.294
6.7.64	3	1	6	2		5		18	1		3		17		21			39	4	0.294
10.7.64	2	1	6	2		5		17	1		3		17		21			38	8	0.294
18.7.64	2	1	6	2		5		16	1		3		17		21			37	2	0.235
20.7.64	2	1	6	2		4		16			3		17		20			36	5	0.250
25.7.64	2	1	6	2		4		16	1		3		16		20		1	37	10	0.250
4.8.64	2	1	6	2		4		16	1		3		16		20		2	38	1	0.250
																			374 Total	

I Infant, *J* juvenile, *Sub* subadult, *Ad* adult, *?* age class unknown.

mean error rate in the last 20 censuses compared with the first 20. Second, there was a small but fairly consistent tendency to under-estimate the number of adults; that is, an individual adult was more likely to be missed than counted twice. Third, the mean error rate, in terms of numbers of individuals, is nearly the same for adult males as for adult females. However, since adult females greatly outnumber adult males in this group (table III) we were more likely to miss a given male than a given female. (This is probably a reflection of the fact that adult males are more likely than adult females to be 'outriders' during group progressions.) It follows that the percentage sampling error is greater for adult males than for adult females, as shown in figure 5.

Infants and young juveniles are not so easy as adults to recognize individually. Consequently it is more difficult to establish the true distribution of their age classes independently of the censuses them-selves. Although we could recognize several of these young animals, there was never a time in our study when we could recognize all of them. Thus, our check on observer reliability for these animals is not on a par with that for adults.

Another indication of the increased accuracy of population data that comes when a group is censused repeatedly is the steady decrease in the percentage of individuals in the census whose sex cannot be determined either through direct observations of the genitals or by recognizing them as individuals (excluding here infants that are identi-fied solely on the basis of being carried by a recognizable mother). This decrease in 'individuals of unknown sex', primarily infants and juveniles, is shown at the bottom of figure 5.

Finally, we note that those censuses which in the field had im-pressed us as being most complete and accurate (as indicated by low error factor and a high evaluation of the quality of the age-sex de-termination) usually were in fact so, as indicated by their close match to the independently-known group composition.

3. Population Statistics, Amboseli

During September, 1963, while deciding upon a group to adopt for concentrated study in Amboseli, we censused several groups. There-after, censuses on Amboseli groups other than the Main Group were taken opportunistically, as time permitted. After eliminating group

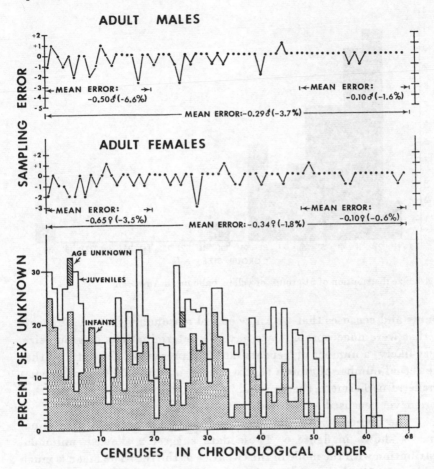

Fig. 5. Census accuracy, based on repeated censuses on the Main Group. Sampling errors for adult males and females are given as deviations of observed values from true values at the time of each census. For juveniles and infants, the lower graph gives the percentage of individuals whose sex could not be determined in the field, either by direct observation of the genitals, or through individual recognition during the census combined with genital observation at other times. Data for juveniles and for animals of undetermined age are given above those for infants, so that the upper line of the graph gives the total number of baboons of undetermined sex. Note the increased accuracy of counts (for adults) and decreased number of animals of undetermined sex that result from repeated censuses of one group.

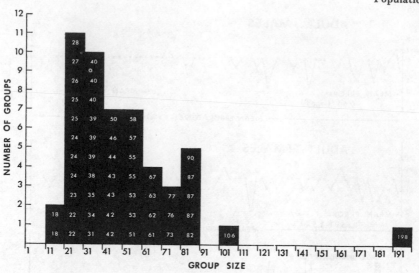

Fig. 6. Size distribution of 51 groups of yellow baboons in Amboseli.

counts and censuses that we know or feel reasonably sure were repeats or that were unacceptable[6], we have 51 determinations of group size. Very likely, a number of 'repeats' are still present in the data, so that the actual number of groups that were censused probably is somewhat smaller. Furthermore, there probably were some groups in the area that were never censused or counted.

The frequency distribution of group sizes in these 51 determinations is shown in figure 6. These data suggest a skewed, unimodal distribution with the mode in the twenties or thirties. The mean is much higher: viz. 51.4 baboons, based on the ungrouped data. These data are

[6] Population data were unacceptable if (1) they included a 'major' inconsistency, e. g. the census data on one group added up to 65 baboons, yet at least 78 individuals were counted; (2) the census or count is known or suspected to contain 'large' errors of omission or commission, e. g. because one subgroup did not pass the counting point, because some individuals ran past the counting point too rapidly to be censused or counted, or because youngsters ran back and forth across the counting point so rapidly that we could not keep track of them; or (3) if the technique that was used was poor, e. g. the count was taken in a quick scan of the group, the counting point did not provide a sufficiently clear view, or the record is ambiguous (as in the expression 'juvenile-two or young adult female'). In more intensive work on primate populations, criteria of adequacies for censuses should be made more explicit than we have done, e. g. precisely how large can inconsistencies get before the results are unacceptable?

comparable to those obtained in Amboseli by WASHBURN in 1959, based on 15 group counts [DEVORE and HALL, 1965] (see table VII).

Of these 51 determinations of group size, 20 are based upon censuses in which the age or sex of at least 89% of the members of the group were recorded. Again, we have eliminated those censuses that are known to be 'repeats' or that were otherwise unacceptable. Census data for these 20 groups are given in table V. In sum, the data indicate a population with 19.3% infants, 23.3% juveniles, 30.3% adult females, 4.1% subadult males and 22.9% adult males.

The overall sex ratio among the adults was 0.76 males per female, that is, about 4 females to every 3 males. But differences between groups were considerable. The sex ratios in these 20 groups, as well as in all others in the literature on baboons, except hamadryas, are depicted

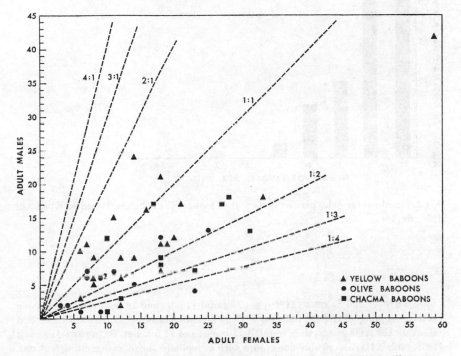

Fig. 7. Baboon sex ratios, based on our own observations and on the following literature: yellow baboons, this study; olive baboons, HALL [1965b], DEVORE and HALL [1965] and MAXIM and BUETTNER-JANUSCH [1963]; chacma baboons, HALL [1963] and STOLTZ and SAAYMAN [1969]. Data from MAXIM and BUETTNER-JANUSCH [1963] have been plotted as olive baboons, but the animals may be cynocephalus baboons (see footnote 2).

in figure 7. These data may be compared with those compiled by TOKUDA [1961–62, fig. 1] for several other species of primates. In the one-male hamadryas units, the sex ratio depends entirely upon the number of females. The distribution of adult females in hamadryas units is shown in figure 8.

Particularly noteworthy in our data is the occurrence of groups with more adult males than adult females. This has previously been reported only once[7] in baboons—in a group of anubis baboons in Queen Elizabeth Park, Uganda [ROWELL, 1964]—and only infrequently in other species of primates.

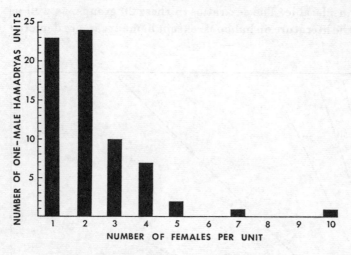

Fig. 8. Number of females per one-male unit of hamadryas baboons. Based on KUMMER, 1968; fig. 14.

[7] STOLTZ and SAAYMAN [1969] got 12 [adult] males and 10 [adult] females in one census on a group of chacma baboons in the Northern Transvaal, S. Africa. In a second census on the same group, they counted 11 males and 11 females. BUETTNER-JANUSCH [1965, table XIII] lists two baboon groups with a preponderance of males, citing DEVORE [1962] as the source. But to judge by the original [DEVORE, ibid., fig. 24], it appears that the sex ratios quoted by BUETTNER-JANUSCH from DEVORE's data are based on animals of all ages, not just adults. Contrary to BUETTNER-JANUSCH [ibid., No. 8 and 9], DEVORE's data do not include a group of 38 with a sex ratio of 1.2:1, nor a group of 33 with a sex ratio of 1:2.

4. Population Dynamics, Amboseli

During our study, the size and composition of the Main Group changed as a result of five processes: birth, death, emigration, immigration, and maturation. No permanent splitting of the group or permanent amalgamation with another group was observed. All known changes in the Main Group's composition are indicated in table IV. As a check on completeness, if one begins with the initial composition of the group and runs through all of these changes, the result will be the composition of the group at the end of our study; the compositions at these times are given in table III.

In actual practice, we found it useful to write out the changes in the opposite direction. We began with the composition at the end of the study, when our ability to recognize individuals was greatest and our censuses most accurate, and worked our way backwards, checking results against the 68 complete censuses on which figure 5 is based. Only one discrepancy was found that was not noted in the field and that is not attributable to census error: a juvenile-1 male was missing in all censuses after June 24. That juvenile has been included in table III and IV. All other changes in table III and IV were noted in the field. We believe, therefore, that the list of changes given in table IV is complete, and we are confident that the only possible exception would be individuals that entered and left the group when we were not present, such as an infant that was born and then died during one of our absences from Amboseli. Even this possibility is unlikely during 1964, for by then we recognized each adult female of the group and kept records of the females' sexual cycles. Depigmentation of the paracallosal skin areas (changing from black to pink) occurs during pregnancy; further, these areas turn from pink to brillant carmine near term, then repigment in the first few weeks *post partum* [GILLMAN and GILBERT, 1946]. No female of the Main Group went through all of these changes during 1964, but was not observed with an infant. This leaves only the possibility of an animal that migrated into the Main Group while we were away, then out again before we returned, or vice versa.

Birth dates in the Main Group (fig. 9) suggest that births in Amboseli may be seasonal, but much larger samples are needed before we can be confident of this.

The data in table IV indicate that during the 372-day period from 28 July 63 until 4 August 64, the Main Group gained 13 members

Fig. 9. Distribution of births in Main Group.

Table IV. Vital statistics for Main Group

No.	Date	Gains				Losses				Maturations						
		Event	Age	Sex	Name	Event	Age	Sex	Name	Age	Sex	Name	→	Age	Sex	Name
1	7.9.63	birth	infant	♂	of Kink											
2	12.9.63	birth	infant	♂	of Goldy											
3	4.10.63	birth	infant	♂	of ♀3											
4	6.10.63					death	infant-1	♂	of young adult ♀							
5	23.10.63					death	infant-1	♂	of Kink							
6	18.11.63	immig.	adult	♂	Shag											
7	18.11.63	immig.	adult	♂	Whitetip											
8	1.1.64									juv.-2	♂	3	→	subad.	♂	3
9	1.1.64									inf.-2	♂	Whitey	→	juv.-1	♂	Whitey
10	1.1.64									inf.-1	♂	of Goldy	→	Inf.-2	♂	of Goldy
11	1.1.64									inf.-1	♂	of ♀3	→	inf.-2	♂	of 3
12	1.1.64									inf.-2	♀		→	the small juv.-1 ♀		
13	12.1.64	birth	infant	?	of Shorty											
14	16.2.64	immig.	adult	♂	Newman											
15	7.3.64					emig.	adult	♂	Kink							
16	12.3.64					death	infant-1	?	of Shorty							
17	18.3.64					emig.	adult	♂	Humprump							
18	19.4.64					death (?)	adult	♀	Curvaceous							
19	23.4.64					death	adult	♂	Three							
20	23.4.64					death	juv.-2	♂	Blacky							
21	15.5.64					death	infant-2	♂	of ♀3							
22	25.5.64	birth	infant	♀	of Notch											
23	25.5.64	birth	infant	♂	of Round Hips											
24	30.5.64 to 4.6.64					death (?)	adult	♂	Newman							
25	5.6.64	birth	infant	♂	of Arch											
26	22.6.64 to 25.6.64					death (?)	juv.-1	♂	none							
27	3.7.64					emig.	adult	♂	Even Steven							
28	6.7.64	birth	infant	♂	of ♀15											
29	10.7.64					death	infant-1	♀	of Notch							
30	18.7.64					death (?)	adult	♂	BBT							
31	20.7.64	birth	infant	?	of Concave											
32	25.7.64	birth	infant	?	of Kink											
33	4.8.64					death (?)	adult	?	Calico							

(10 births, immigration of three adult males) and lost 15 (12 known or suspected deaths, emigration of three adult males) for a net loss of two.

The number of days during the study that the group maintained each size is shown in figure 10. Although these data give the impression of coming from a population with a simple, unimodal distribution, such as would be expected if there is an equilibrium group size, a distribution like this might also result from a small sample of a random walk process that would depend upon the size of the group at the beginning of the study period.

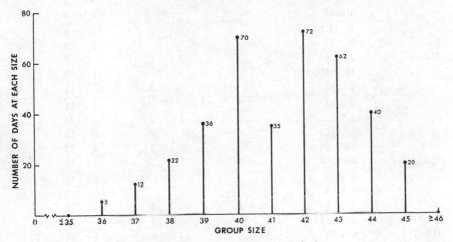

Fig. 10. Number of days that the Main Group spent at each size.

Perhaps a more revealing way of looking at size distribution data is suggested by figure 11. If group size tends to increase when small and decrease when large, then the points above the midline in such a graph will tend to lie to the left of those below the line. No such trend is apparent, suggesting that change in group size may be independent of the size of the group before the change, but the lack of obvious trend may be the result of small sample size.

During the course of our study, the sex ratio (males per female) among the adults of the Main Group varied from 0.555 to 0.235. Thus, the number of adult females per male varied from less than two to more than four.

These changes in sex ratio took place as a result solely of deaths and migrations, the latter being restricted to adult males: only adult

Table V. Composition of groups of yellow baboons in Amboseli, based on 20 groups (of the 51 used for figure 4) in which the age or sex of at least 89 % of the members of the group were recorded

Group name	Census date	Males								
		I_1	I_2	I_{1-2}	J_1	J_2	J_{1-2}	Sub	Ad	?
Small	29.3.64							1	3	
—	25.10.63								2	
—	5.8.63							1	10	
—	9.9.63					2		3	5	
—	4.8.63							2	6	
—	29.11.63					2		1	6	
—	7.9.63							1	9	
Main	29.7.63	1	5		3	3			7	
—	12.9.63				3			1	9	
—	2.8.63					5		1	11	
—	5.8.63							2	12	
—	9.9.63			1	1	2		2	24	
—	31.7.63				1	2		2	11	
TTF	29.3.64				2	5		2	17	
Shaggy's?	29.9.64		2		1	7		4	11	
—	6.9.63							3	16	
—	3.8.63							8	9	
—	7.9.63		1			2		1	18	
—	25.7.63					11		2	21	
HB	6.9.63					6		8	42	
Totals		1	8	1	11	47	0	45	249	0
% of total		0.08	0.71	0.08	0.98	4.2		4.0	22	0

I Infant, I_{1-2} infant $_1$ or infant $_2$, *J* juvenile, J_{1-2} juvenile $_1$ or juvenile $_2$, *Sub* subadult, *Ad* adult, *?* age class unknown.

males migrated into or out of our main study group. These males ranged in age from the youngest fully mature male in the group to the oldest. Specifically, three males, Shag, Whitetip and Newman, moved into the group, whereas three other males, Kink, Humprump, and Even Steven, moved out. Shag and Whitetip entered the Main Group together; Newman, about three months later. What group any of these males came from is unknown. Kink and Humprump both moved into a group called TTF (for 'twisty-tailed female'), but at different times.

Table V. (continued)

males							Sex unknown								Total	Error factor	
I_2	I_{1-2}	J_1	J_2	J_{1-2}	Ad	?	I_1	I_2	I_{1-2}	J_1	J_2	J_{1-2}	Ad	?		+	–
					6		2	1		5					18	1	0
			1		12		1	1	1						18	–	–
					6		1	3		1					22	–	–
					8		1	2		2	1				24	2	2
					8		1	4	2	2	4				29	4	4
					12			7		1				2	31	5	0
					12		1	4	1	1	5			4	38	8	2
1·				2	18				–						40	0	0
					14		1	4	1	4	5				42	0	0
					7		2	8		5	2			3	44	5	–
					20		2	12	1	3	1				53	1	1
			1		14			6	1	1					53	2	0
		1	3		18		1	4		2	7			3	55	–	–
		1	1		21		1	3		8					61	–	–
		1			19			13	1	1	2				62	4	1
					16		3	7		8	7			3	63	0	0
					8			19	6	6	11		1	5	73	–	–
					33		4	8		11	5		4	3	90	6	5
			1		18			12	13	18	10				106	–	–
					59		5	24	3	16	23	5	2	5	198	15	3
1	0	6	6		329	0	26	142	30	95	83	5	7	28	1120		
0.08	0	0.53	0.53		29	0	2.3	13	2.7	8.5	7.4	0.44	0.62	2.5			

Male Kink's migration will be described in detail below (p. 47). Male Even Steven moved into Fang's group (named for a male with a protruding canine). Male Newman was last seen in the Main Group on May 5, 1964; we do not know whether he died or emigrated.

These data are in marked contrast to those obtained in 1959 by DEVORE for olive baboons in Nairobi Park, Kenya. During the 10 months of his study on 5 groups, only one animal, an adult male, is known to have changed groups [DEVORE, 1962; DEVORE and HALL,

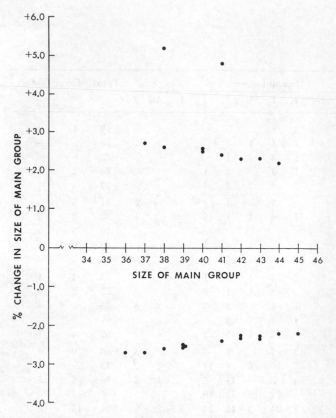

Fig. 11. Relationship between percent change in size of Main Group and the size of the group before the change. A point is plotted for every day on which there was such a change, as shown in table III. If there is an equilibrium group size, then the size of the group will tend to increase (+) if the group is below this size, and to decrease (—) if above it. The present small sample would be adequate to reveal such a trend only if it were very marked.

1965]. Another adult male, in a different locale, was apparently observed to change groups by WASHBURN and DEVORE [1961], who indicate a total of two migrants.

Gains and losses in adult males are of particular interest. It may be that adult males are more likely to leave or die in those groups that have a relatively larger number of males, and more likely to enter those in which adult females are relatively numerous. If so, such intergroup migrations would tend to maintain a stable sex ratio. By compensating for small-sample anomalies in sex ratios at birth, they would be of particular significance in small groups.

In figure 12, the number of days that the Main Group spent at each sex ratio is plotted. As in the case of change in group size, the question of an equilibrium distribution may be approached better by considering the percent change observed at each group size. Figure 13 plots this relationship for changes in the sex ratio of the Main Group. As before, no trend toward an equilibrium was revealed by the available sample.

Of the other groups in the Amboseli study area, one of the most fascinating is HB Group, so called because of the presence of a hunchbacked adult male (fig. 14). In November, 1963, this group, with 198 animals, was one of the largest groups of wild nonhuman primates ever recorded, exceeded in size only by the nightly aggregations of hamadryas units at sleeping rocks [KUMMER, 1968] and by hand-fed groups of Japanese macaques on Mt. Takasaki, one of which included over 700 monkeys in October, 1967 [ITANI, 1967].

On 9 separate days during our study, we censused or counted the members of HB Group. The results are shown in table VI. In looking at these data, the reader should bear in mind that a group of this size

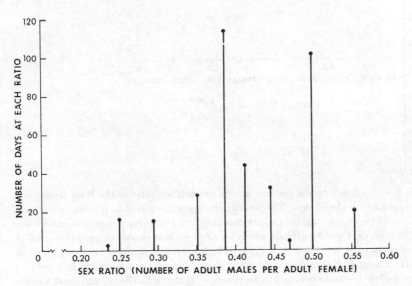

Fig. 12. Number of days that the Main Group spent at each sex ratio.

Table VI. Changes in composition of HB Group, Amboseli

No.	Date	Males						Sub	Ad	?
		I_1	I_2	I_{1-2}	J_1	J_2	J_{1-2}			
1	26. 7. 63								2	
2	31. 7. 63					1		6	36	
3	6. 8. 63							12	60	
4	6. 9. 63					6		8	42	
5	14. 9. 63					1		8	35	
6	15. 9. 63					2		1	3	
7	12. 10. 63								5	
8	3. 11. 63									
9	8. 5. 64								1	

Abbreviations as in table V. The two subgroups shown for September 15 were separated for 2 ½ h. Other counts on the first subgroup: 63, 64, 64, 65, 65, 67. Other count on the second subgroup: 119.

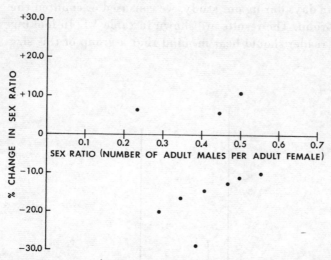

Fig. 13. Relationship between percent change in adult sex ratio of the Main Group and the sex ratio before the change. The percent change is calculated as follows: 100 × new sex ratio — old sex ratio ÷ old sex ratio. A point is plotted for every day on which there was such a change, i. e. for every day on which an adult male or female entered (matured or immigrated into) or left (died or emigrated from) the group, using the data given in table III. If there is an equilibrium sex ratio, then the sex ratio will tend to increase (+) if the group is below this ratio, and to decrease (—) if above it. The present small sample would be adequate to reveal such a trend only if it were very marked.

Table VI. (continued)

I_2	I_{1-2}	J_1	J_2	J_{1-2}	Ad	?	I_1	I_2	I_{1-2}	J_1	J_2	J_{1-2}	Ad	?	Total	+	−
					1									176	179	10	3
	1				26		1	29	7	14	15			48	185	6	2
			1		34			9	9	12	20			22	180	15	4
					59		5	24	3	16	23	5	2	5	198	15	3
					37		1	17	1	7	10			1	118	2	2
					2									59	67	(188)	
														121	121		
					1									172	178	2	2
															193	193	
					1					2				167	171	5	5

is extremely difficult to census or count accurately, and that the figures in table VI are based on single counts or censuses, rather than repeated observations on each day. Thus, while some of the apparent changes in this group doubtless are indicative of population dynamics, others may be the result of not seeing some subgroups, or of other errors in observation.

Although this immense group often moved as a unit, it sometimes split into subgroups. On September 14, 1963, only 118 members were found together. Next morning the group was together again, but later that morning we watched as the group split again into two large subgroups (table VI, No. 6). These two subgroups were observed for 2½ h after the split at the end of which time they amalgamated again into a single group.

A group of 171 baboons and one of 185 were observed in Amboseli in 1959 by WASHBURN [DEVORE and WASHBURN, 1963]. Unfortunately, there is no way to know whether either was the antecedant of HB Group.

These censuses of groups in Amboseli are probably not complete, even for groups whose home ranges overlap that of our Main Group, and there are doubtless inaccuracies in those censuses that are presented, to judge by fluctuations observed in results from repeated censuses on several groups. The data are presented both for comparison with earlier

14

15

Fig. 14. The hunchbacked adult male of HB Group.
Fig. 15. Isolated juvenile with paralyzed hind legs.

work on baboon populations and in the hope that they may stimulate further research in this area. We feel strongly that a study of the population dynamics of Amboseli baboons, based upon repeated censuses of known groups over a long period, is an unusual research opportunity. We know of no other population of nonhuman primates, in a relatively undisturbed natural habitat, that provides better opportunities for direct observations on population dynamics and for relating these processes to underlying social and ecological factors.

5. Yellow Baboons Outside Amboseli

Yellow or cynocephalus baboons were encountered in only one place during our travels outside Amboseli: a group of well over 80 animals was observed at Olnaigon Swamp, near Kimana, Kenya (fig. 1). VAN CITTERS and his colleagues [1967] counted the baboons in six groups in this area. The counts were 8, 13, 39, 58, 79, and 'at least 150'. The composition of three groups of baboons near Masalani, Kenya, were given by MAXIM and BUETTNER-JANUSCH [1963] and are shown in our table VII. As indicated earlier (fn., p. 4), these are probably yellow baboons. We do not know of any other census work on yellow baboons outside Amboseli.

6. Other Baboon Species

Our counts and censuses on groups of olive (anubis) baboons in several locations are given in table VIII, along with counts and censuses from the literature on this species. Population data for chacma baboons are compiled in table IX[8]. For hamadryas baboons, the distribution of females (adult, subadult and juvenile) per one-male unit is shown in figure 8 [based on figure 14 in KUMMER, 1968]. KUMMER writes: 'At their peaks, hamadryas units contain from 2 to 5 females, most of which are adults' [ibid., p. 34]. These units, along with several other

---

[8] Some compilations [e.g. HALL, 1963, fig. 1; DEVORE and HALL, 1965, fig. 2–4] of HALL's chacma group counts disagree with the raw data [e.g. HALL, 1963, fig. 2] and with each other. We have used the most specific raw data in the literature, as cited in our table IX.

Table VII. Size and composition of yellow baboon groups, based on literature

Location, date	Male									Female							
	I_1	I_2	I_{1-2}	J_1	J_2	J_{1-2}	Sub	Ad	?	I_1	I_2	I_{1-2}	J_1	J_2	J_{1-2}	Ad	?
Masalani, Kenya, 1960?							8	4								23	
Masalani, Kenya, 1960?							2									4	
Masalani, Kenya, 1960?							7									11	
Amboseli, Kenya, 1959																	
Kimana, Kenya, (date ?)																	

kinds of 'detached parties' [ibid., p.86], coalesce into large troops each evening at the sleeping cliffs. The composition of hamadryas troops may vary from night to night, however, and, 'the smallest independent entity within the troop may well be the one-male unit' [ibid., p.99].

7. Isolated Baboons

Although baboons usually remain in obvious association with their group, they may become isolated at times. In the literature, social isolation has been treated as an all-or-none affair, but our observations on Amboseli baboons indicate that the extent and duration of isolation can vary greatly. These differences are important for ecological and behavioral reasons.

We distinguished a number of degrees of isolation from a group. At a minimum, an individual may be spatially separated from his group, though still within view and hearing range. Of course, this kind of quasi-isolation occurs constantly, in that there is always some member of the group who is farther from his nearest neighbor than is any other member of the group. As an example of an extreme of such separation, adult Male Three slept separately one night, but within visual and auditory range of the Main Group.

Table VII. (continued)

unknown						Totals	Reference
I_2	I_{1-2} J_1	J_2	J_{1-2} Ad	?			
10		15				60	MAXIM and BUETTNER-JANUSCH, 1963
		2				8	MAXIM and BUETTNER-JANUSCH, 1963
3		4				25	MAXIM and BUETTNER-JANUSCH, 1963
						13, 42, 47, 51, 57, 64, 66, 70, 74, 78, 88, 94, 103, 171, 185	DEVORE and WASHBURN, 1963
						8, 13, 39, 58, 79, $\geq$150	VAN CITTERS et al., 1967

Next, an animal may be temporarily out of sight of other members of his group but still within hearing range of it. This is a common phenomenon, particularly when the group is moving through areas of dense foliage. A low-pitched, quiet grunting (the 'cohesion' grunt) apparently enables animals to maintain auditory contact during such times. Such mild isolation was too commonplace to record.

Third, from time to time during a progression, an animal may be out of sight and sound of any other member of his group but continue to orient toward them. For example, there sometimes are 'outriders' during progressions, particularly when the group is in the widely-dispersed formation that is characteristic of foraging periods. Occasional glimpses of the group, combined with a knowledge of a progression route, probably enable such animals to continue to move parallel with the group and subsequently to rejoin it. Such outriders are usually adult males. Shag, an adult male who migrated into the Main Group during our study, was particularly prone to being an outrider.

Another major cause of temporary isolation without loss of orientation is the capture of prey (to be discussed in detail below). At such times, an adult male that is eating meat, and sometimes one or more other males that sit near him, 'waiting' for scraps, will lag far behind their progressing group, to the extent that the group may go completely

Table VIII. Size and composition of anubis baboon groups, based on literature and this study

Location	Date	Group name	Males								
			I_1	I_2	I_{1-2}	J_1	J_2	J_{1-2}	Sub	Ad	?
Nairobi Park, Kenya	1959	KR									
Nairobi Park, Kenya	1959	MR								1	
Nairobi Park, Kenya	1963	MR			3	5	2			4	
Nairobi Park, Kenya	1959	AR								1	
Nairobi Park, Kenya	1963	AR									
Nairobi Park, Kenya	1959	LT			2	3	3			2	
Nairobi Park, Kenya	1963	LT									
Nairobi Park, Kenya	1959	SR			2	5	1			6	
Nairobi Park, Kenya	1963	SR									
Nairobi Park, Kenya	1959	SV				8	6			5	
Nairobi Park, Kenya	1963	SV									
Nairobi Park, Kenya	1959	KV									
Nairobi Park, Kenya	1963	KV				11	6			13	
Nairobi Park, Kenya	1959	HP									
Nairobi Park, Kenya	1959	PP									
Nairobi Park, Kenya	1963	PP									7
Murchison Falls, Uganda	1963										6
Murchison Falls, Uganda	1963										6
Murchison Falls, Uganda	1963										7
Murchison Falls, Uganda	1963										12
Murchison Falls, Uganda	1963										
Murchison Falls, Uganda	1963										
Murchison Falls, Uganda	1963										
Waza Reserve, Cameroon	1967?										12
Nairobi Park, Kenya	1963	SV					1		2	8	
Nairobi Park, Kenya	1963									5	
Lake Manyara Park, Tanzania	1963					1	2		5	13	
Lake Manyara Park, Tanzania	1963						1		1	4	
Ngorongoro Crater, Tanzania	1963									1	
Seronera, Serengeti, Tanzania	1963									4	
Seronera, Serengeti, Tanzania	1963									9	

Table VIII. (continued)

I₂	I₁₋₂	J₁	J₂	J₁₋₂	Ad	?	I₁	I₂	I₁₋₂	J₁	J₂	J₁₋₂	Ad	?	Total	Reference
															61	DeVore, 1965
					6				5						12	DeVore and Hall, 1965
	1	2	1		6										24	Hall and DeVore, 1965
					9				18						28	DeVore and Hall, 1965
															39	DeVore, 1965
	1	1		3			2								17	DeVore and Hall, 1965
															27	DeVore, 1965
	4	3		7			3								31	DeVore and Hall, 1965
															40	DeVore, 1965
1	3	3		14			2								42	DeVore and Hall, 1965
															34	DeVore, 1965
															77	DeVore, 1965
	1	9	6	25					3	2					76	Hall and DeVore, 1965
															87	DeVore, 1965
															24	DeVore, 1965
															47	DeVore, 1965
					7				4						18	Hall, 1965b
					9				4						19	Hall, 1965b
					9				9						24	Hall, 1965b
														21	28	Hall, 1965b
					18?				8					10	48	Hall, 1965b
															14	Hall, 1965b
															42	Hall, 1965b
															46	Hall, 1965a
														25	37	Struhsaker, pers. com.
					5			1	3	5	5				30	this study
					2		1	2	1	3	2	2		21	39	this study
			1		17			1	7		6			28	81	this study
		1			11		1	3	2	3	3			21	51	this study
					1		1	1						4–6	8–10	this study
					2				1	1				2	10	this study
					2				1	1		1		20	34	this study

Table IX. Size and composition of chacma baboon groups, based on literature

Location, date	Male						Female					
	I_1	I_2	$I_{1-2} J_1$	J_2	J_{1-2} Sub Ad	?	I_1	I_2	$I_{1-2} J_1$	J_2	J_{1-2}	Ad
Cape of Good Hope Nature Reserve, S. Africa												
1960		2		2?	2				3		9?	8
1961	3	3		5	2		3	2			6	11
1958–1959					1							10?
1958–1959					5							21
1961					8							18
Table Mt., S. Africa 1959					3							12
N. Transvaal, S. Africa												
21.7.67					17							27
12.1.68					18							31
24.7.68					18							31
9.10.67					18							28
29.2.68					19							24
26.8.68					15							25
10.10.67					9							18
17.8.68					9							15
8.8.68					12							10
21.8.68					11							11
9.10.67					7							23
23.8.68					9							20
9.10.67					17							17
20.8.68					16							21
Mana Pools, Rhodesia, 1961												
Giant's Castle Reserve, Drakensburgs, 1961												
Windhoek, S. W. Africa, 1960												
South Africa* (date ?)												
Windhoek, S. W. Africa, 1960												

The size of HALL's troops marked with an asterisk were obtained by subtracting HALL's data for the Ca
Reserve, Table Mt. and Giant's Castle Reserve (shown below) from the total distribution shown in HA
1963, fig. 2, S. Africa.

Table IX. (continued)

I₂	I₁₋₂	J₁	J₂	J₁₋₂	Ad	?	Group name	Totals	Reference
							S	26	HALL, 1962a
							S	35	WINGFIELD, in HALL, 1963
		9					N	20	HALL, 1963
		27					C	53	HALL, 1963
		54					C	80	HALL, 1963
	13							28	HALL, 1963
3		17					W	68	STOLTZ and SAAYMAN, 1969
2		12					W	72	STOLTZ and SAAYMAN, 1969
3		15					W	77	STOLTZ and SAAYMAN, 1969
		10					RB	57	STOLTZ and SAAYMAN, 1969
1		7					RB	60	STOLTZ and SAAYMAN, 1969
9		8					RB	59	STOLTZ and SAAYMAN, 1969
		4					KMO	37	STOLTZ and SAAYMAN, 1969
5		4					KMO	36	STOLTZ and SAAYMAN, 1969
		2					K	24	STOLTZ and SAAYMAN, 1969
6		2					K	30	STOLTZ and SAAYMAN, 1969
		5					KMW	41	STOLTZ and SAAYMAN, 1969
		3					KMW	36	STOLTZ and SAAYMAN, 1969
		5					GK	41	STOLTZ and SAAYMAN, 1969
1		6					GK	45	STOLTZ and SAAYMAN, 1969
. . . .								13, 11–15, 21–25, 31–35, 36–40, 36–40, 36–40, 46–50, 51–55, 51–55, 66–70, 86–90, 109	HALL, 1965b
. . . .								15, 21, 21–22, 37, 58	HALL, 1963
. . . .								one group of 1–10 six groups of 10–19 eight groups of 20–29 two groups of 30–39 and one each 40–49 and 50–59	HALL, 1963
. . . .								10–19, 10–19, 30–39, 30–39, 40–49, 40–49	HALL, 1963
. . . .								65	HALL and DeVORE, 1965

out of sight. We have circumstantial evidence that an adult male baboon (not from the Main Group) that was eaten by lions was one who caught a hare, then lagged behind his group as he sat and ate it.

Fourth, an animal may become temporarily lost from its group. Animals in such a situation are clearly distinguishable by their persistent searching and visual scanning activities. Two adult females (one from the Main Group), three adult males (one from the Main Group), and one large juvenile male of the Main Group were observed temporarily lost from their groups, each on a separate occasion. The juvenile male, Drape, became separated from the Main Group as they moved through tall, thick grass. Drape headed toward another group, looking quite confused, i.e., frequently starting, stopping and looking around. He was vigorously attacked by several adult males of the other group, but eventually managed to escape. After about 45 min of being separated from the Main Group, he was back in it again, licking his bloody wounds.

The final two categories consist of animals that are completely isolated from any social group and show no orientation, searching, or scanning activity. These include, first of all, animals that are physically capable of keeping up with a group but who are not, at least at the time, members of any group. Seven such adult males, and one pair consisting of an adult male and an adult female, were seen. The latter two approached but did not join the Main Group. In each of these 8 cases, we could not find any other group for which these baboons might have been 'outriders', and we are fairly confident that all of them were isolates.

The first isolated male was at one of the largest temporary rain-pools (No. 40, fig. 38) when the Main Group arrived. He approached Male Five, the most dominant male of the group, who was seated at the time, and tugged at Five's hip—apparently incipient mounting. Male Five remained seated for a while, then quietly walked away. The lone male was then approached by a large infant, who stood with its arm outstretched to the male, either 'milk-begging' or showing incipient grooming. The male remained immobile, sitting quietly. The male did not join the group as they left the rainpool. He was not seen subsequently, either near the group or at the rainpool.

A second isolated adult male that approached the Main Group was chased off by a large juvenile male of the group. Two other isolated males watched or were watched by members of the Main Group, but did not join the Group.

Of the three remaining isolated males, one had a large, bloody wound on his right leg when first observed, but over one month later, when observed again, had no obvious impairment, yet had still not rejoined a group. The other two had no visible wound or impairment. One of these two males, as well as the wounded male, exhibited hiding and 'freezing' behavior when approached by us, behavior that we have not seen in any other adult males, except occasionally in an adult male who migrated from our main study group to another, and whose behavior during the transition period was observed (Male Kink, see below).

Last, we come to animals that are separated from a group for physical rather than social reasons: they are physically incapable of keeping up with a moving group of baboons. In this class were a crippled, small juvenile (fig. 15) and, on another occasion, an adult female with her accompanying large infant. The adult female moved very slowly, with her eyelids half closed. Members of the Main Group, particularly two adult females, watched this mother and infant for some time but the latter two made no attempt to join the group.

Isolated baboons have seldom been reported in the literature. HALL [1963] wrote that, 'only one probably genuine isolate,... a near-adult male, has ever been observed by us (in S. Rhodesia), although a full-grown male was reported by the Cape Reserve staff to be regularly seen during the 1961 observation period there'. WASHBURN and DEVORE [1961] saw only 3 solitary baboons, two females and one male, all of which were either very sick or badly hurt and trying to rejoin a troop.

One striking conclusion is suggested by these observations: unless physically incapacitated (or in the case of a dependent infant, unless associated with an incapacitated mother) animals other than adult males are never totally isolated from a social group.

8. Migratory Males

(a) Male Kink

As mentioned above, Male Kink was one of the adult males that migrated from the Main Group to another group in the area. He was an animal that we could readily recognize and that could be observed at close range with a minimum of disturbance. Because his migration presented a unique opportunity to witness the process of intergroup

migration, we watched Kink whenever time allowed during this period.
A résumé of significant events is given below.

While Male Kink was not one of the most sexually active members
of the group, we saw him mount adult females on 9 occasions before
March 4, 1964, including at least 4 complete mountings, i. e. terminating
in intravaginal ejaculation. At times, he seemed to be a somewhat
peripheral member of the group, both spatially and socially.

March 4, 1964 (Day 1), at 1633. Kink chased an adult female (Stocky), in low estrus,
away from adult male Newman, and later walked up to her and mounted. *1635.* After a
fight with adult males ES and BBT, he chased an adult female in middle estrus.

Day 3, 0818–1230. Kink mounted adult female Stocky 5 times during this period. He
ejaculated during at least the first mount. Between mountings, he persistently followed
her at an unusually close distance. It was 7 days before the onset of her detumescence:
she was in the follicular phase of her menstrual cycle.

Day 4, 0851–0927. Kink continued to consort with Stocky and mounted her at least
once during this period. When the dominant male of the group approached them while
they were grooming each other, they stopped grooming and gave an incipient avoidance
movement. When the dominant male continued past them, they sat down again and
resumed grooming.

Day 6. Male Kink could not be found in the Main Group.

Day 7, 0907. The dominant male of the Main Group mounted female Stocky and ejaculated. *0924.* Kink was located ¼ mi. north of the Main Group, progressing parallel to
them. There was no indication of tension or anxiety on his part. *0940.* The dominant
male of the Main Group moved 50 ft. away from the group, approaching within 50 ft. of
Kink, who was seated and feeding at the time. Perhaps in response to this, Kink ran,
passing through a gap separating the Main Group from another group, which was
progressing immediately behind it. The dominant male of the Main Group then returned
to his group.

From then until at least 1051, when we lost sight of him, Kink followed
another group (probably the TTF Group). During this period, he walked
behind or parallel to them and remained 75–80 ft. from the nearest
individuals, with no apparent interaction. Kink alternated between
foraging on the ground and looking around him from vantage points
in fever trees.

Day 8, 0928. An adult male not from the Main Group (possibly from the TTF Group)
chased Male Kink, who ran up a tree. Another male from the same group ran up a nearby

tree and bellowed. Male Kink remained in the tree until 0942, at which time he was approached by adult male Whitetip of the Main Group and his consort female. Male Kink descended the tree and walked away from the Main Group.

Except for a few short periods when foliage obscured our view, we kept Kink under continuous observation for the rest of the day. Until 1658, he was not involved in any further interactions with groups. He foraged, progressed, and repeatedly scanned the area, utilizing 11 different trees as vantage points, from which he could see various baboon groups in the area. Doubtless, it is not by chance that he was not in the immediate vicinity of any of them.

1658. After 91 min in a tree from which the Main Group was visible, Male Kink descended and slowly approached the Main Group, walking carefully. When Kink was 60 ft. from the group, the most dominant male left the group and approached him at a walk. Kink fled, and disappeared from sight 300 yards away. The dominant male walked alone in the same direction for about ¼ mi., then sat on a stump 5 ft. off the ground, watching in the direction in which Kink had vanished. Subsequently, the dominant male climbed a fever tree, from which he looked about in the direction that Kink had taken, and then, at 1714, he went up a snag, from which he bellowed.

When the dominant male left the group, many other members of the group poured out of the grove of trees that they had been under and moved 60 ft. out toward the male, then stopped, forming a 'pseudopod', much like that found in hamadryas baboons under similar circumstances [KUMMER, 1968, p. 112].

1720. The dominant male moved back toward the Main Group.

1727. We found Male Kink about ½ mi. away, in a tree. He looked northeast toward an approaching group, then descended and walked westward. During the next 54 minutes, he progressed fairly directly to Sleepy Hollow, the most frequently used sleeping grove of the Main Group. No group was near him as he did so. He ascended a tree in Sleepy Hollow at 1821. Two groups (but not the Main Group) slept within ³/₁₀ mi. of him. No other baboons slept in Sleepy Hollow that night.

The next morning (day 9), Kink was located at 0731 h, alone and asleep in the trees of Sleepy Hollow. He did not descend until 0911 at which time ejaculate could be seen dripping from his fully extended penis.

0929–1029. Kink progressed steadily, with several stops for visual scanning from trees or from a bipedal stance. He followed the same route that he had used on the morning of day 7. He walked past water in a rainpool without drinking, ran through a treeless area, and then vanished from sight at 1029.

1030. At the approach of an observer, Kink was suddenly flushed from dense foliage and ran away rapidly and continuously, then vanished again in undergrowth. The place where he was last seen was approached by an unknown adult male, whose gait and watchfulness were strikingly like those displayed the day before by the dominant male of the Main Group, as was the fact that this male was approximately ½ mi. away from his presumed group. This male then climbed a fever tree and watched in the direction in which Male Kink had vanished. The male remained until at least 1045. A search by the observer for Male Kink during the next hour was unsuccessful.

Day 10, 1719–1740. A search for Kink during this period was unsuccessful. He was not seen in or near the Main Group on this day.

Day 11. Male Kink was not seen.

Day 12. No observations were made on this day.

Day 13, 0942. Kink was found, alone and on the ground, in the vicinity of trees in which the Main Group and two others (HB Group, White Infant Group) had slept. We observed him until 1111. At 1000, he approached and was 'greeted' by an adult male, Male Six, of the Main Group: as Kink walked past, Male Six momentarily touched his muzzle to Kink's side. Kink ignored this. He remained in the vicinity of the Main Group until 1020, then moved progressively farther away from them, foraging as he went and occasionally scanning the area, either from a tree or bipedally from the ground.

Day 14, 0906. Kink was found on the ground, in the vicinity of six other groups, including the Main Group. At 0926, he was approached by a young adult male (Even Steven) of the Main Group while the dominant male watched. Kink moved away from them, approached an adult male of another group and grabbed at that male's pelvis (a 'greeting'). The male walked past Kink, approached and was then groomed by an adult female. When Kink approached them, the male pushed closer to his female, but she walked away when Kink got within 5 ft. The male then turned on Kink with jaws open and bellowed at him. Kink ran a short distance away. The male returned to his female companion, and they resumed grooming.

0946. A second approach and greeting by Kink to this male and female resulted in another threat from the male, who slashed at Kink with open jaws, and then hit at him. Kink walked away. As he did so, he looked back toward the Main Group and toward the group to which the male belonged. When Kink moved away, the dominant male of the Main Group, who had moved away from his group toward Kink, turned and walked back to his group.

1043. Two adult males (not from the Main Group) approached Kink. One walked past Kink. The other touched muzzles with Kink twice. Kink walked away.

From then until 1120, Kink interacted with these two males and two others, always peripheral to any group. The interactions included a variety of behaviors that might be called 'greetings'. Thereafter until 1215, Kink remained near the group from which these males came (TTF Group ?). During this period, he foraged, progressed, and scanned the other group from a dead tree.

We next saw Kink at 1743, within the TTF Group from which he ran 60 ft. At the time, the TTF Group was being closely followed by the Main Group. During the next hour, Kink had several interactions with adult males of the TTF Group, most of which were 'greetings' (e. g., the males touching the genitals of Kink; muzzle contact, fig. 16). On several occasions during the hour, Kink was completely surrounded by baboons of the TTF Group. Thus, almost two weeks after leaving the Main Group, he was physically within another group. At 1845, he moved at the front of the TTF Group—as though leading them—when they moved toward and up a grove of sleeping trees.

It is interesting that this was a grove of trees used by the TTF Group, but not by the Main Group. We could not tell whether Kink's 'choice' was based on anticipating the line of march of the TTF Group or whether his behavior was based on long familiarity with this group and its customary sleeping groves.

Male Kink was seen in the TTF Group again the next day, and on 22 other days between then and the end of our study. Indeed, he was never again seen outside that group. He ignored an estrous female that presented to him on March 22 (day 19), but mounted a female on June 27 (day 118). He was a well-integrated member of the TTF Group, even during agonistic encounters between that group and the Main Group. His behavior and social relations seemed quite indistinguishable from those of a naturalborn member of the TTF Group. To determine whether that is generally the case would require a more thorough study of migrant animals than has so far been made.

There were several noteworthy events during Kink's transition from the Main Group to the TTF Group. He left shortly after consorting with an estrous female who later consorted with the most dominant male; this was probably no coincidence. He was prevented from return-ing to the Main Group by that male, and, while we did not see what happened on the day of Kink's departure, it seems likely that this male was instrumental in evicting him.

Next, we note that Kink was able to survive for nearly two weeks in isolation from any group. His one known sleeping grove during this period was the favorite sleeping grove of his former group. Yet he quickly adopted the grove of his new group once he was accepted by them.

While Kink occasionally hid in undergrowth, like other solitary animals, he was generally very vigilant, frequently scanning the area either bipedally from the ground, or from vantage points in trees.

He was often on the periphery of other groups, where his interactions were virtually all with other adult males. These interactions were usually any of a variety of behaviors that we tentatively class as 'greetings'. His eventual access to an estrous female of the TTF Group indicates the role of migratory males in genetic exchange between groups.

Finally, we note that the group which he joined is one with which the Main Group interacts perhaps more than any other. Furthermore, it is the group that adult male Humprump joined. Humprump was last seen in the Main Group on March 18 (day 15 of Kink's migration) and was next seen on June 22, in the TTF Group, at which time he, Kink, and a young adult female were the three lead members of the group.

On the basis of migrations of other males into or out of the Main Group, we can add further to the picture of migratory males, although for none of these other males is our data as complete as it is for Kink.

(b) Newman

Male Newman, who had moved into the Main Group about February 16, 1964, was last seen in the group on May 30 of that year. The most noteworthy event in the weeks preceding his exit was an intense and prolonged display of aggression by the dominant male against him and male Whitetip on May 23. The agonistic encounter began with an attack by the dominant male on Whitetip, but the latter quickly established a coalition with Newman. Despite synchronized attacks on the dominant male by Whitetip and Newman, the two of them were clearly defeated by the dominant male.

On the evening of the 30th, the last day we saw Newman, he lagged 80 ft. behind the group as they moved toward their sleeping grove. Subsequently, when Whitetip was again attacked by the domi-

16

17

Fig. 16. Muzzle contact of an adult male of TTF Group with Male Kink from the Main Group.
Fig. 17. Male Even Steven (left) pulling back scalp of male Humprump. Note Even Steven's hand on Humprump's head. Adult Male BBT is in foreground.

nant male and tried again to establish a coalition with Newman, Newman simply walked away.

Newman was never seen in another group. We do not know whether or not he moved into one.

(c) BBT

Male BBT (bouncy, black tail), like Newman, was simply missing one day (July 18, 1964) and was never seen subsequently. During the preceding months he was not involved in any outstanding social interactions; indeed, he seemed to be a rather non-interactive male. But he became partially paralyzed early in 1964, as did a number of other baboons in Amboseli. We first picked up symptoms of paralysis in BBT on April 20[9]. By the end of May, he was walking with his toes dragging in the dust and without fully extending the legs, so that the pelvis was carried lower than normal. The gait was slow; the pelvic region and lower extremities seemed stiff. These symptoms are typical of other cases that we saw[10]. On June 29, we noted that BBT's paralyses was getting worse. On July 13, when the rest of the group, at the edge of a waterhole, became alarmed and suddenly took flight, BBT remained behind. Perhaps he crouched and froze, as we have observed in young animals. He remained in the Main Group until July 17, but was never seen again.

While male Newman may have made a successful transition to another group, it seems unlikely to us that BBT could have survived alone during the transition period.

(d) Humprump

Male Humprump was a large, fully adult male, about 15 years old. He was one of the most subordinate adult males of the Main Group. On

[9] We have one note made October 29 of the previous year, that BBT limped and that his legs seemed stiff.

[10] On the basis of serological studies at NIH in one such paralyzed baboon from Amboseli and in our son, there is reason to suspect that a cocksackie virus, perhaps B_2 or B_4 was spreading through Amboseli at the time, and may have been responsible for the paralyses [c.f. KALTER, RATNER, RODRIGUEZ and KALTER, 1967].

February 29, Humprump consorted with a female that had a large sexual-skin swelling (i.e., that was in estrus), and they apparently mated. Later that day, there were extensive and intense agonistic interactions among several males of the Main Group, but because of rain, we could see little of what transpired. The next day, three adult males (Even Steven, BBT, and Male Six) ganged up on Humprump in one of the most persistent and intense coalition displays that we saw. Humprump was eventually reduced to a state of waxy paralysis and sat that way despite repeated bites and threats, even when Even Steven sat behind him, pulling back his scalp (fig. 17). For the next few days, Humprump was harassed by Even Steven and BBT. He was last seen in the Main Group on March 18, 1964, and when found 96 days later, in the TTF Group, he and Male Kink were among the lead members in a progression.

Finally, we note here the possibility that adult male Humprump was temporarily in another group on November 3, 1963. He was out of, or very peripheral to, the Main Group from the afternoon of October 18 until November 5, except for November 2. This period was fairly early in our work on the Main Group, and our ability to recognize individuals was not yet well developed.

(e) Even Steven

Adult male Even Steven himself later left the Main Group. On June 12, the dominant male chased adult female Null away from him; at the time, Null (for nulliparous) was the female of the Main Group with the largest sexual-skin swelling. (Ten days later, her sexual skin detumesced.) Thirteen days after this chase, Even Steven gradually moved ahead of the Main Group during a progression and moved over to another, larger group. But he was chased all the way back to the Main Group by an adult male of that group. Two days after that, he was attacked by the dominant male of the Main Group. On July 2 he consorted with and mounted an elderly female, Calico, who was in estrus (six days prior to detumescence). Later that day, however, he was again attacked by the dominant male, who had mated with this female on June 30, and did so again on July 2, 4 and 5. The following day was his last day in the Main Group. Six days thereafter he was located in Fang's Group.

(f) Immigration

So much for the adult males that left the Main Group. What happens, subsequently, when a male attempts to enter another group ? We have presented our observations on Male Kink. Transitions of the other males that left the Main Group were not observed, but we did make observations on three adult males, Shag, Whitetip, and Newman, after they entered the Main Group.

Shag and Whitetip arrived together, suggesting that they may have come out of the same group. Furthermore, they were often closely associated in the Main Group. They were first noticed in the Main Group during a census, on November 18, 1964. (On November 3, a male that was probably Shag had been seen in another group.) Like Male Kink, they were often at the rear of progressions during the early days of their move into their adopted group.

Shag, Whitetip and Newman were all attacked, quite forcefully, by the dominant male of the Main Group during the months following their entrance into the group. As indicated above, Newman left again, after 104 days in the group. Shag and Whitetip were still in the group when our study terminated, 260 days after they moved in.

Newman, the immigrant male that later left the group, was only once observed to mount a female and then he apparently did not ejaculate (no ejaculatory pause and no ejaculate visible externally). In other respects, Newman seemed to be one of the most subordinate adult males of the group. In contrast, Shag and Whitetip completed copulations with several adult females of the Main Group.

When these cases of migratory males are reviewed, a common pattern seems to emerge. The emigrant male had lost a fight—either with the most dominant male of the group or with a coalition of other males—several days before his departure. Yet in none of the observed cases were there any conspicuous wounds. In at least one case, competition for estrous females seemed to be involved. After a transition period of several days, during which the migratory male was independent of any group, he began to move into a new group, at first following it at a distance. Early encounters were primarily with adult males of the new group; these relations with 'outriders' may be essential to the eventual acceptance of the newcomer. A variety of behaviors that might be called 'greetings'—perhaps because we do not yet recognize their true function—were common during the early days in the new group.

The most subordinate of the three males that entered the Main Group never, to our knowledge, mated successfully with any adult female of the group, and thus did not contribute to genetic interchange between groups. He eventually left the group. In contrast, each of the other two immigrant males mated with several estrous females and, so far as we know, remained in the group.

Finally, some of our observations suggest that new males are eventually singled out for attack by the dominant male of the new group, though we are not yet sure that such fights are more common than those between the dominant male and other males of the group.

9. Comparisons and Discussion

(a) Group Size

The variability of baboon group sizes within any one area is striking. The variability recorded in Amboseli appears to be the most extreme case: group sizes ranged from 13 to 185 in Washburn's sample[11] and from 16 to 198 in ours. Furthermore, there seem to be marked differences in mean group size from one area to another. In Amboseli, mean group size was 51.4 during our study; in Nairobi during DeVore's study, a sample of 9 groups had an average of 42; and in South Africa the 15 groups that Hall censused had a mean size of 31.

Commenting on such differences, DeVore and Hall [1965] wrote that,

'these data from Kenya, where the groups of 12[12] and 185, for example, shared the same area, indicate that group size in an area is related more to patterns of social behavior than to differences in local environment. The only suggestion that local ecological conditions may influence group size comes from the single study area where differences in altitude were marked. In the Drakensberg Mountains of South Africa a brief sampling of the population showed a negative correlation between the height of a group's home range and the number of individuals in the group.... It is not unreasonable to suppose that group size in this area is dependent upon the food resources. While baboon foods are relatively plentiful at the lower elevations, they steadily decrease in the upper reaches. Because they move as a group as they forage for food, there may be an

[11, 12] The size of the smallest group censused in Amboseli by Washburn during 1959 was reported as 13 by DeVore [1962], but as 12 by DeVore and Hall [1965].

upper limit to group size in areas where food is scarce. Baboons usually travel about three miles a day, and a large group would have more difficulty accommodating the nutritional needs of all its members when food is hard to obtain.'

Relating group size to social behavior and to local environment are not alternatives: they are two different aspects of the same phenomenon. The coexistence of groups of different sizes in the same area, even with partially or totally overlapping home ranges, does not rule out the possibility that there is an optimal group size for any particular set of environmental conditions such that the net reproductive rate for groups of that size will be greater than the reproductive rate of groups of any other size. If such is the case, then groups whose sizes are near the optimum will grow at the expense of those that are more aberrant. Another possibility is that groups of different sizes occupy somewhat different niches[13].

KUMMER [in DeVORE, 1965] writes:

'Hamadryas baboons sleep exclusively on vertical rocks. Now, such rocks are very numerous in the western areas, while they are scarce in the east. On the other hand, agriculture, that is, food, has an inverse distribution. We found a highly significant increase of troop size from the west toward the east, which probably means that the dense population supported by the rich food supply in the east is forced together in large numbers on the few rocks available.'

By 'troop' KUMMER refers not to the one-male social units, but to aggregations of them at the sleeping rocks. In hamadryas, however, the foraging unit is not the troop, but the single male with his associated adult females and offspring. These one-male units are considerably smaller than most social groups of savannah baboons, and KUMMER himself [in KUMMER, 1968, p.155] speculates that

'whereas sparseness of food seems to favor a population with small, dispersed social units, the scarceness of groves or cliffs for sleeping requires exactly the opposite, i.e., large concentrations of animals in a few spots.... The hamadryas apparently has specialized to alternately meet both needs. Its bands can split up into one-male units which may disperse during the day in search of food. On the other hand, bands can develop a relationship that permits them to share a sleeping rock, i.e., to form a troop.'

[13] ROWELL [1966, p.360] has speculated that even within a baboon troop, competition may be reduced by age-specific dietary preferences.

When supplemental food was given over many years to rhesus monkeys *(Macaca mulatta)* on Cayo Santiago and to Japanese macaques *(M. fuscata)* on Mt. Takasaki and elsewhere in Japan, group sizes increased steadily. This suggests that the natural abundance of food may be a limiting factor affecting group size. '...It would seem that the determinants of regional variations in group size are primarily to be found in food density as averaged over a year' [HALL, 1965].

The relationship of group size to food abundance in savannah baboons is not obvious. A complicating factor and one that may be of particular importance in establishing optimal group size is the influence of visibility on the coordination of group movements. In the open, low grassland of Amboseli, visibility is unusually good, particularly in the foraging areas, where groups are usually most widely dispersed. The terrain is virtually flat and the grass in most places is quite low. Over a comparable distance in Nairobi Park, baboons in one part of the group would more often be hidden from those in another by the higher foliage or by irregularity in terrain. On the other hand, Nairobi Park, with its high rainfall, probably provides a richer food supply for the baboons. Yet the minimum, maximum and mean group sizes are smaller in Nairobi Park than in Amboseli.

Whatever the ecological determinants of group size, they must operate through those population processes, such as birth and death, that affect group size. Estimates of the rates of these processes in baboons are given below. Their effects on the distribution of group size will be discussed in chapter IX.

(b) Birth Rate

Some data are available with which to estimate the birth rate for female baboons. During our study of yellow baboons, 10 infants were born in the Main Group (table IV) during a total of 6,608 female-days (obtained from data in table III). This gives a reproductive rate of 1.513×10^{-3} births per female per day, or an average of one infant per female every 661 days. ROWELL [1966, table III] gives the number of adult females in two groups of olive baboons as 21^{14}. From January, 1963, through

---

[14] According to ROWELL [ibid., p. 351], some of these females reached adulthood during the study, but no more specific data are given about them. Counting these females as adults throughout the study tends to deflate the estimated reproductive rate.

March, 1965, 26 new infants were observed in these two groups [ibid., table IV]. Assuming that all births were seen and that all adult females were present throughout the study, these data yield an estimate of 26 births per 21 females per 824 days, or 1.503×10^{-3} births per female per day, equivalent to one infant per female every 665.5 days, on the average[15]. These two estimates are in close agreement.

DeVore [1962, fig.24] recorded 7 infants born during his study. Thirty-seven adult females, not counting two that disappeared from one group during the study, are listed in his census data on the 5 groups involved [ibid.]. Assuming that DeVore recorded all births in these groups during the 9-month period, April–December, 1959, and that all adult females were present throughout the study, these data give a reproductive rate of 0.69×10^{-3} infants per female per day, or an infant per female every 1,454 days[16]. This last estimate seems unreasonably low. It may be that breeding tends to be seasonal in Nairobi Park baboons, as it seems to be in Amboseli, and that many infants were born during the months that DeVore was not present. Alternatively, a number of births may have been missed.

According to Gilbert and Gillman [1951], the post-partum amenorrhea of lactating chacma baboon mothers lasts 5–10 months. 'In two exceptional mothers...the perineum commenced to turgesce on the 33rd and 78th days...post-partum. Normal menstrual cycles were established immediately despite the fact that the babies were still at the breast.... We have reason to believe that they were ovulating on those cycles.'

Kriewaldt and Hendrickx [1968] indicate that 'successive full-term births after an initial pregnancy occur 277 ± 80 days [range ?], or approximately 6.5 to 12 months apart.' This datum is based on females whose infants were taken from them at birth [Hendrickx, personal communication], and thus may be used as an estimate of interbirth interval in wild females whose infants die shortly after birth.

---

[15] Rowell [ibid., p.353], estimated the interval between births as follows: 6 months (pregnancy) + 5 months (lactation interval) + 5 to 15 weeks (one to three menstrual cycles before pregnancy). This would indicate a birth interval of about 369–439 days.

[16] DeVore [ibid., p.104] estimated the interval between births at twenty to twenty-four months (about 607 to 730 days), as follows: 12 to 15 months in the period of post-partum amenorrhea + 2 or more months until the onset of the next breeding season + 6 months gestation.

Only one female of the Main Group, female Kink, had two infants during our study (table IV). Her 1963 infant died at the age of 46 days, on 23 October 1963. Thereafter, Kink's sexual skin became tumescent by the end of November, and she mated with several males at that time, and during the first week of December[17]. We were away from Amboseli in early January, when her second estrus period was expected[18]. She was again swollen and in estrus during early February (presumably her third cycle); detumescence began on February 14. There were 172 days between the onset of this detumescence and the birth of her next infant[19]. The infant was born 286 days after the birth of her previous infant.

Of the other 8 mothers in our study, 5 showed no sexual skin swelling at any time from the birth of their infant until the termination of our study (periods of 10, 29, 60, 71, and 327 days). The infants of all 5 were still alive at the end of our study. Another female, Shorty, had an infant that died at the age of 60 days. As in the case of female Kink, the death of the infant seems to have initiated the onset of cycling. Her first post-partum swelling detumesced on 24 April 64, which was 43 days after the death of her infant. During her second post-partum cycle, in

[17] KRIEWALDT and HENDRICKX [1968] indicate an average of 46.4 days in the 'recycling periods' of baboons after natural live births. According to HENDRICKX [personal communication] the recycling period is the interval from delivery until the first sign of sexual skin swelling, and the reproductive parameters that are given in the paper cited are for females whose infants were removed from them at the time of birth and were not subsequently returned.

[18] Mean cycle length in chacma baboons is 39.63 days [GILLMAN and GILBERT, 1946]. According to SMITH et al. [1967] the menstrual cycle in baboons (species unspecified) lasted for 35.4 days in caged females without any adult male baboon in their colony, but for only 30.3 days after two adult males were brought in, even though neither male had access to the females. At the same time, the period of 'estrus' (= sexual skin tumescence?) was extended from 3.0 days to 10.83 days.

[19] According to GILLMAN and GILBERT [1946], ovulation precedes detumescence by two to three days. The gestation interval indicated here corresponds closely with the period of gestation that has been observed in captive chacma baboons (187 days, range 173 to 193 days, in 14 cases; GILBERT and GILLMAN, 1951), and hamadryas baboons (172.2 days; ZUCKERMAN, 1932). GILBERT and GILLMAN's intervals are based on timed matings, in which the female is put with the male from 'shortly before the expected time of ovulation, that is, at the time of maximum perineal turgescence, and remained with the male until the first signs of deturgescence, ... usually ... three to four days'. They do not say what they count as day zero of pregnancy, but assuming it is the presumed conception day, their estimates would in general be a few days longer than estimates based on the onset of deturgescence (= detumescence).

which detumescence began on or about June 5, she apparently became pregnant: a slight reddish tinge was noted in the paracallosal skin[20] on July 26 and was more noticable on August 1. The infants of the two remaining mothers died during the study. The infant of Female Three died on 15 May 64 at the age of 194 days. It was still suckling in the weeks preceding its death. The mother had small sexual-skin swellings during mid-June and mid-July. (Such small swellings are common in the first post-partum swellings of baboons: GILLMAN and GILBERT, 1946.) Female Notch's infant died on 10 July 64 at the age of 46 days. On July 22, Notch was noted as having a small sexual-skin swelling.

(c) Death Rate

The crude death rate in our Main Group was 12 deaths (or presumed deaths) in a total of 15,449 monkey-days, or 7.77×10^{-4} deaths per monkey-day. Much of this rate is accounted for by infant mortality. There were 4 deaths during 596 monkey-days for infant-ones (first half-year), equivalent to 6.71×10^{-3} deaths per infant-day. For older infants and juveniles combined, there were three deaths in 5,441 monkey-days, or 5.51×10^{-4} deaths per monkey-day. For adult females, the rate was two deaths in 6,608 monkey-days, or 3.03×10^{-4} deaths per monkey-day. For adult males, the rate was 3 deaths in 2,804 monkey-days, or 1.07×10^{-3} deaths per monkey-day. (Deaths are listed in table IV; monkey-days for any age-sex category were calculated from the data in table III.)

(d) Emigration Rate

As described in this chapter, only adult males were observed to migrate from group to group. There were three emigrations from our Main

[20] There seems to be no standard term in the literature for this area of bare skin. GILBERT and GILLMAN [1951] refer to it as 'the non-oedematous bare areas of the skin lateral to the callosities'. It is present in both males and females, but does not change color in males. It is not part of the sexual skin in the strict sense: it does not swell during the sexual cycles. Depigmentation of this paracallosal skin, with a change from black to pinkish red, is an indication of pregnancy and begins during the second week after the expected menstruation [GILBERT and GILLMAN, 1951].

Group (table IV) in 2,804 monkey-days for adult males (table III), equivalent to 1.07×10^{-3} emigrations per male-day, or an emigration by each male every 935 days, on the average.

(e) Group Fission and Intergroup Migration

Only a few data are available in the literature to indicate the frequency or circumstances under which individuals or groups of individuals leave groups. ROWELL [1966] indicates that two adult males moved into one of her groups of olive baboons in Uganda. In another troop, the number of adult males 'fluctuated between 4 and 5 and once reached 6; two individuals were always present, and a third for most of the time, but the remaining one or two had often changed between sessions...and some older juveniles may have left....' In addition, an adult male vervet monkey *(Cercopithecus aethiops)* lived for at least two years with one of these baboon groups before moving to another one, 'where it was equally accepted in grooming and sexual interactions' [ROWELL, ibid.]. ROWELL [ibid.] continues as follows:

> 'With this amount of movement between troops—the occasional joining together of two troops in foraging expeditions (S and F), and the fact that as the V troop increased in size it occasionally divided into smaller parties which followed different routes during part of the day—the validity of the troop, as a population unit, might be questioned. This is a theoretical query, there was no subjective doubt in the field as to the entity of the troops, probably because each had a stable nucleus of known individuals; in particular there was no indication that any female, and certainly no adult female, ever exchanged troops.'

DEVORE [1962] noted one baboon, an adult male, that changed groups during his 1959 study in Nairobi Park. He presents circumstantial evidence that two of the groups in the park may have split between that time and his 1963 recensus [DEVORE, 1965]. 'The survey also confirmed that group membership remains stable over long periods, even when the population is subjected to extreme drought and flooding' [DEVORE, ibid.]. No further data are given on intergroup migration.

During HALL's studies of chacma baboons in South Africa, and of olive baboons near Murchison Falls, Uganda, he apparently never saw intergroup migration [cf. HALL, 1962a, p.191], although in both areas he did see adult males that seemed to be isolated from any group [HALL, ibid. and 1965b].

MAXIM and BUETTNER-JANUSCH [1963] observed seemingly solitary baboons on four occasions. In three cases, they were large males;
the fourth was an adult female, who—judging by the authors' description of her behavior—probably was temporarily lost.

STOLTZ and SAAYMAN [1969] have not seen any intergroup migration in their current study of chacma baboons in the Northern Transvaal.
They write: 'Variations [over time] in troop size were probably accounted for by the natural increase and mortality of baboons rather
than by the interchange of individuals from one troop to another;
observer error undoubtedly accounted for a small percentage of the
discrepancy.'

(f) Group Composition and the Sex Ratio

Published census data on the age-sex composition of groups of yellow,
olive, and chacma baboons are given in table V, VII, VIII, and IX.
The composition of hamadryas troops, which probably are not homolog-
ous with the groups in savannah baboons, are given in table X. The
number of females per one-male hamadryas unit are given in figure 8.
Accompanying these females are their dependent offspring.

Table X. Percent composition of hamadryas groups at Erer-Gota, Ethiopia, based on
KUMMER, 1968, table VII

	Adult	Sub-adult	3½–2½ years	2½–1½ years	1½–½ years	½–0 year	Total
Males	22.9	6.0	4.3	4.9	6.5	1.9	46.5
Females		32.6	5.6	6.1	7.5	1.8	53.6
Total		61.5	9.9	11.0	14.0	3.7	100.1

On the basis of these data on baboon populations, several tentative
generalizations can be made about the composition of baboon social
groups. So far as is known, none is devoid of preadolescent animals. All
contain at least one adult male and one or more adult females. Furthermore, one-male units are characteristic only of hamadryas baboons;

multi-male groups are virtually universal in olive, yellow, and chacma baboons. Except for one group of olive baboons, the yellow baboon is the only species in which groups have been reported with more adult males than adult females.

In a baboon group, unequal numbers of adult males and females may result from any of several factors. First, there may be systematic differences at birth. Although we do not yet have enough records on the sexes of baboon neonates to rule this out, major differences seem unlikely, on the basis of records from other primates (e. g., GILBERT and GILLMAN [1951] reported the births of 7 male and 7 female chacma baboons; VAN WAGENEN [1954] reported 103 male rhesus to 105 female rhesus born in her colony; the Yerkes colony has yielded 151 live-born chimps to date, of which 75 were male) as well as on theoretical grounds [FISHER, 1930].

Secondly, if sex determination is independent from birth to birth, small-sample differences in sex ratios of infants will be almost inevitable among small groups. For example, 7 of the 8 infants born in our Main Group during this study are of known sex, and of these 7, 6 are males. While this would seem to be a highly disparate sex ratio, getting at least 6 infants of the same sex out of 7 infants can be expected in about 13 out of every 100 cases, even if, at birth, the chance of a male is one-half. In any case, the result is that variability in infant sex ratios will be greater in small groups. This, in turn, may have marked effects upon behavior and social structure, as ROWELL [1966, p. 351] has emphasized[21]. For example, what becomes of an infant female whose only available playmates are male? What happens in a group when a new crop of juveniles, almost all of whom are males, reaches sexual maturity?

Third, disparate sex ratios of adults may result from differences in maturation rates, hence differences in the ages at which males and females are called 'adults'. Commenting on this, DEVORE and HALL [1965] write:

> 'One of the most important reasons for the disparate sex ratio in adult baboons is the fact that females mature in roughly half the time that males do. By the time she is three and one-half or four years of age the female is coming into estrus. By the age of five she is bearing young and has reached full physical growth.

[21] The model of independence between group composition and the probabilities of behavior in animals of each age-sex category is given in ALTMANN [1968].

The young male, although he is sexually mature by five, takes from seven to ten years to attain full physical growth...an observer counting baboons, then, will classify as "adult" all females with infants, or in estrus, or above a certain size. "Adult males", however, are usually only those which have attained the full measure of maturity.'

An implication of this explanation is that there should be an excess of older juvenile males.

Two other factors that may affect the adult sex ratio of groups are differential rates of migration (in the narrow sense of moving into and out of social groups) and differential mortality. As we have indicated above, migration of baboons in Amboseli may be a major factor affecting adult sex ratios of individual groups, but intergroup migration alone can have no effect on the overall sex ratio in the population of groups, except insofar as migratory males go through a solitary period during which they are not counted as contributing to the sex ratio of any group. It is noteworthy that, except for incapacitated animals, only adult males were observed living in isolation from any social group, and that only adult males are known to migrate from one group to another. As for mortality, the data from our Main Group suggest a higher mortality in males: the data in table III and IV indicate 1.916×10^{-3} deaths per monkey-day for males, but only 0.256×10^{-3} deaths per monkey-day for females.

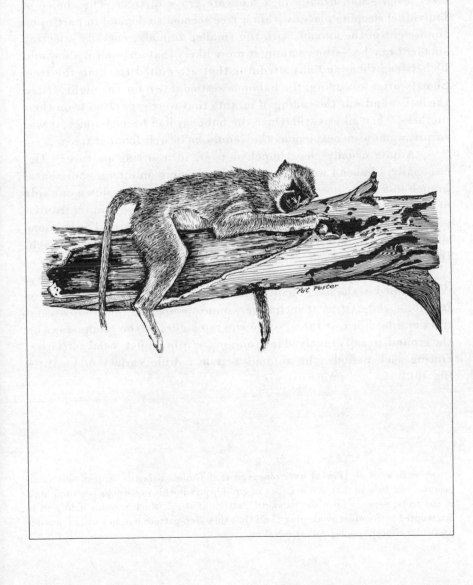

In the evening, at dusk, the baboons of Amboseli climbed into fever trees, each group usually in a separate grove of trees. The choice of individual sleeping places within a tree seemed to depend in part upon the weight of the animal, with the smaller animals generally selecting smaller branches—thus making it more likely that an adult male would be between them and any predator that attempted to climb the tree. Shortly after ascending, the baboons settled down for the night. Often the last sound was the calling of infants that were separated from their mothers. Then all was still. Once the baboons had become quiet, it was surprising how inconspicuous they could be in well-foliated trees.

Adults usually slept hunched over, in a sitting posture [22]. Occasionally a baboon was seen resting or sleeping on a large, horizontal branch in a prone position, the left arm and leg hanging down one side of the branch, the right arm and leg down the other (see chapter frontispiece). Small infants slept in their mother's ventral flexure. The baboons sometimes seemed to get into such a position, in the fork of a tree branch, that they would not fall, even without muscle tonus.

Baboons also rest or even nap from time to time during the day. As described in the next chapter, many afternoon progressions slow or stop for a while. Often at such times, some members of the group ascend trees and feed or rest there, while the remainder of the group stays on the ground, usually involved in grooming or other quiet social activities. During such periods, the animals rest in a wide variety of postures (fig. 18).

[22] BERT et al. [1967a] have observed that Guinea baboons sleep in this same posture, and believe that maintenance of equilibrium in this precarious position, high in the trees, depends upon the baboons' pattern of sleep, which remains light and is interrupted by frequent awakenings, and that this sleep pattern has, in turn, required a particular neurophysiological organization.

Fig. 18. Resting postures of baboons. (a) An adult male in an unusual, cat-like posture. (b) A napping adult male being groomed by a large infant.

c

d

Fig. 18. Resting postures of baboons. (c) A large juvenile female in a common resting posture. (d) A subadult male napping in a braced position, with fingers held under foot.

1. Sleeping Groves

The night-time sleeping grove of the Main Group is known for a total of 125 days. The locations of the groves that were used are shown in figure 19. The frequency with which each grove was used is shown in table XI. Table XII gives the frequencies of transitions from one grove to another for 88 pairs of consecutive nights.

By a *sleeping grove* we mean a cluster of trees that are sufficiently close together that the baboons can get to any tree within it from any other tree in it without descending to the ground, and that are surrounded by gaps too wide to be crossed without descending[23].

The Main Group had a 'favorite' grove (grove 1, fig. 19), which accounted for 32% of their night locations[24], and which we nick-named 'Sleepy Hollow'. If all members of the Main Group slept in Sleepy Hollow any one night, there was about a 48 percent chance (13/27) that they would return the next night. Thus, the fact that the group slept in Sleepy Hollow the night before made it even more likely that they would return. The next-most-popular grove, grove 2, accounted for 25% (32/125) of their sleeping sites, and they slept in that grove 28% (7/25) of the nights in which they had slept in that grove the night before. Thus, this grove did not appear to exhibit the kind of residual attraction in the choice of sleeping site that grove 1 did. All of the remaining groves together accounted for 42% (53/125) of the Main Group's sleeping sites, yet for these groves there were only nine percent repeats, all accounted for by grove 3. Thus, it would appear that the remaining groves show, if anything, a negative residual influence; that is, the probability that the Main Group will sleep there on any given night is lower if they slept there the previous night. [Excluded from this analysis are a case in which the group was divided between groves 2a and 1 on one night, then all slept in grove 1 the next night, and another case in which after

[23] We believe, but are not certain, that groves 2a and 2b (fig. 16) are, in this sense, distinct groves; although data for 2a and 2b have been tabulated separately, they have been lumped in the analyses described below. Similarly, we believe but are not sure that grove 1 consists of two separate groves, which have been lumped in our tabulations and calculations. On the other hand, groves 11a and 11b are probably separate groves, and they have been kept separate in the analysis.

[24] 35% if we include four nights in which the group was divided between grove 1 and another grove (table XII).

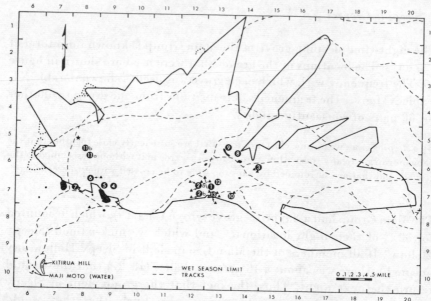

Fig. 19. Sleeping groves of the Main Group (numbers in black circles) and of other groups (black triangles). The solid line is the outer boundary of the home range of the Main Group. Dotted lines (west side of range) indicate positions of the outermost members of the group when the group was near the periphery of its range. Dash lines indicate vehicle tracks. The star indicates the position of a grove of trees that was almost adopted by the Main Group. Groves 4–7 and 11 are in KH woods; the others are in KB woods.

Table XI. Number of nights during which baboons of the Main Group used each of their sleeping groves

Grove No.	Frequency of use	Grove No.	Frequency of use
1	40	10	1
2a	15	11a	1
2b	16	11b	4
3	13	12	0
4	4	13	1
5	9	1 and 2a	3
6	5	1 and 12	1
7	1	2a and 2b	1
8	3	4 and 5	2
9	3	5 and 6	2

The locations of these groves are shown in figure 19. Where more than one grove number is given, the group split into subgroups, each of which slept in a separate grove. (See footnote, p. 71, for some possible errors in grove demarcation.)

Table XII. Frequencies of transitions by the Main Group from one grove to another on successive nights

First night \ Next night	1	2a	2b	3	4	5	6	7	8	9	10	11a	11b	12	13	12+1	4+5	5+6	2a+2b	2a+1	Total
1	13	2	1	3		4			1							1		1		1	27
2a	6	2	1	3																	12
2b	3	2	1	1		2							1							1	12
3	4		3	3																	9
4	1																				1
5	1		2		2	1											1				8
6							2	1					1								4
7												1									1
8	2																				2
9	2				1																3
10					1																1
11a																					0
11b	1												1					1			3
12																					0
13	1	1																			1
12+1																					0
4+5																					0
5+6																			1		1
2a+2b			1																1		1
2a+1			1						1												2
Total	33	7	10	8	4	7	2	1	1	1	1	1	3	0	1	1	2	1	1	3	88

dividing between these two groves one night, they all slept in grove 2b
on the next night (table XII)].

These residual effects cannot be accounted for by proximity to the
grove of trees in which the animals last slept. As we show in the next
chapter, they usually ranged, during the day, far outside the woodland
areas in which the sleeping groves were concentrated, and when return-
ing the next evening, would often pass up several groves in their progression toward the grove of their 'choice'.

Altogether, there were 27 of 88 days (30.6% with 95% c.l. 22–42%)
during which the Main Group progressed in a 'closed loop', i.e. some of
the group returning in the evening to the same grove from which members of the group had descended that morning. Of these 27 days, there
were 23 on which all members of the group returned to the one grove
in which they had slept the night before.

On a few occasions, the Main Group was observed to leave one
grove of trees in the evening and move to another. This was almost
always the result of intergroup competition for sleeping trees.

When such competition took place late in the evening, it sometimes resulted in a group split, some members of the group going off to
another grove, others staying in the contested grove. Intergroup competition was not always present during a division of the group among
sleeping groves, however: we have observed such a split in the Main
Group when no other group was near them. Members of the Main Group
slept in two separate sleeping groves on eight nights out of 125 (6.4%,
with 95% binomial c.l. 3–13%). On all other nights, all members of the
group slept in but one grove. When more than one grove was used,
there was usually conspicuous ambivalence in the movements of the
baboons as they approached the sleeping groves and at such times some
baboons, even after having partly ascended, descending again and
switched groves. When the group split into such subgroups, the two
sleeping parties were usually both quite large; the composition in one
case is given in table XIII. At the other extreme, adult Male Three slept
alone one night[25].

---

[25] Mathematical models of primate subgroup formation have been presented by
ALTMANN [1965] and, particularly, by COHEN [1968]. The latter has provided a two-
parameter model that fits the distribution of sleeping subgroups observed by STRUH-
SAKER [1965] in Amboseli vervet monkeys, as well as the distribution of (sub)group sizes
in several other primates.

Table XIII. Composition of subgroups of the Main Group on night June 29, 1964

Baboons in grove 5	Baboons in grove 4
Subadult Male 3	Adult female Whip
Adult female Blondy	Adult male Whitetip
Juvenile-1 male of Blondy	Adult Male 5
Juvenile-2 male Drape	Adult Male 6
Adult female Surrey	Adult female Arch
Juvenile-1 male Teddy Bear	Infant-1 male of Arch
Adult female Scyth	Adult Female 3
Juvenile-1 male Whitey	Adult female Calico
Adult female Goldy	Adult female Concave
Infant-2 male of Goldy	Adult male Even Steven
Juvenile-1 male	Adult Female 15
Juvenile-1 male	Adult female Round Hips
Adult male Shag	Infant-1 male of Round Hips
Adult female Notch	Adult female Shorty
Infant-1 female of Notch	Adult female Stocky
Adult male BBT	Small Juvenile-1 Female
	Medium Juvenile-1 Female
	Large Juvenile-1 Female
	Adult female Null
	Adult female Light-Tip
	Juvenile-2 Male 4
	Juvenile-1 male
	Adult female Kink

     The sleeping groves of the Main Group were in two distinct wood-
land areas that are separated by an area with no permanent water and
relatively few trees. We call these wooded areas the KH and KB woods,
for their proximity to Kitirua Hill and Kitirua Beacon, respectively.
If one pools the sleeping tree transition data in table XII into these
woods the following results are obtained:

	Next night	
	KH	KB
First night KH	9	9
First night KB	12	58

from which we get the following relative frequencies:

$$P = \begin{bmatrix} 9/18 & 9/18 \\ 12/70 & 58/70 \end{bmatrix}$$

Using these frequencies as probability estimates, transitions between KH and KB sleeping groves can be represented by a first-order, two-state Markov process:

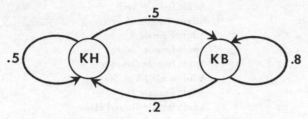

Fig. 20. Relative frequencies of transition between sleeping in the KH woods and the KB woods.

Since P is a regular stochastic matrix, its powers approach a matrix T, each row of which is the unique fixed-point probability vector of P [KEMENY, SNELL and THOMPSON, 1957]. For the values given above for P, that vector is (0.2553, 0.7446). What this means is that the group will spend about 25.5 % of their nights in the KH woods, 74.5 % in the KB woods, and that this distribution is independent of where they were when we began sampling their sleeping groves[26].

Two other estimates of these probabilities can be made. First, the column totals of table XII indicate that 76.1 % (= 67/88) of the nights were spent in the KB woods, 23.9 % (= 21/88) of the nights were spent in the KH woods. COHEN [1968] has proved that this and the previous estimate are virtually identical. Second, a larger sample, including not only those nights for which the next night's sleeping grove is known, but all others as well, is shown in table XI. This second sample gives 22.4 % (= 28/125) of nights in KH woods, 77.6 % (= 987/125) in KB woods. There is close agreement between this estimate and the other two.

Although the baboons sometimes ascended the sleeping trees as soon as they reached the grove, they often coalesced on the ground first. At such times, the group was generally more compact than at any other

---

[26] We have assumed a stationary source, i.e. that the probabilities that we are attempting to estimate are not changing during the period of observation. As has been shown by COHEN [1968], the stationarity of the source cannot be demonstrated from the same set of data as are used to estimate the transition probabilities.

time of day (p.111), and the baboons went through a period of heightened social interaction in which grooming, nursing and other quiet activities were usually predominant (p.93). In some cases the animals seemed obviously fatigued at that time of the day.

The method by which the group's sleeping grove of the night is chosen is not well understood. It will be discussed in a future publication, in which the internal dynamics of the group will be considered. Some factors that may influence the decision are discussed below.

2. Competition for Sleeping Groves

Sleepy Hollow was in an area in which sleeping groves of many groups of baboons were concentrated (fig. 19)[27]. On some evenings, six or more groups, containing over 400 baboons, slept in this area (approx. 0.3×0.6 mi.). As numerous groups converged on this area each evening and then left it again the next morning it became the center of much group interaction. These intergroup relations will be discussed in detail in a subsequent report. They ranged from the virtual ignoring of other groups, to massive and overt aggression. Perhaps the most frequent group interaction was 'pushing', in which one group, with no aggression more serious than approaches and stares, repeatedly or continually induced a group that was ahead of it to move along toward the sleeping groves. Such group supplantation near sleeping trees sometimes took on a tandem formation, with several groups moving along, one behind the other, each spread out in a long file, and sometimes with the front individual of one group and the last individual of the group ahead closer to each other than either was to the nearest member of its own group. (As mentioned above, group censuses at such times may be inaccurate unless one is working with groups of recognizable individuals.)

Occasionally, when more than one group moved into the same grove of trees, there was considerable agonistic behavior between the groups. Doubtless one function of preferred sleeping groves for each group is to reduce the chances of such competition.

We have evidence that the favorite groves of the various groups do not coincide. Sleepy Hollow was seldom used by other groups, even

[27] According to PETER WARSHALL [personal communication] there may be a similar cluster of sleeping groves in a circular patch of trees in the western half of quadrat (11, 7).

when the Main Group slept elsewhere. The Main Group shared the
Hollow with another group (of 18 baboons) on only one night. Similarly,
there were many sleeping groves near Sleepy Hollow that were fre-
quented by other groups, but that were seldom or never used by the
Main Group.

Once the baboons settled down for the night, they apparently
never change groves: on 47 mornings, we found the Main Group in the
same grove in which we had left them the night before, never elsewhere
(95 % binomial c.l. 0.00–0.08).

3. Adoption and Desertion of Groves

Near the end of our study, the Main Group seemed on the verge of
adopting a new sleeping grove, which was on the periphery of the KH
woods, though well within their home range[28] (fig. 19).

On the evening of July 22, the Main Group moved toward the potential new grove from
the east. Six members of the group ascended. Suddenly, most if not all of the baboons
that were still on the ground ran about 50 ft. westward, toward the center of the grove.
They then stopped and looked back. No alarm barks were given by any of the baboons:
the flight of the baboons had apparently been a false alarm. Two minutes later, a mature
adult female (Stocky) initiated a progression of the group out of this grove and to grove
11b, about ¼ mi. away.

The next evening, the group again went to the new grove. Again, the group
suddenly left the area, but in this case, their departure did not begin with an alarm
flight. We did not see who initiated this progression; it was lead by a mature adult
female (Goldy). The monkeys quickly moved eastward entirely out of the KH woods,
then across an open plain to the KB woods. This progression was long, tense and rapid.
(By this time of day, it is getting dark and the baboons are usually in or near their
sleeping grove.) Between 1730 and 1800, the group progressed farther (1.39 mi.) than
during any other half-hour period during our study, for an average of 2.8 mi./h.

Often during this progression, virtually every member of the group,
other than clinging infants, was running. Repeatedly, the progression
stopped, and numerous baboons, starting with those near the front,
stood bipedally, looking sometimes to the north, sometimes south, but
in each case, virtually all looking in the same direction. The subadult

[28] On 118 nights before that, the Main Group slept elsewhere (binomial confidence
limits 0.00–0.03). Furthermore, this grove was not among those used on 125 nights in the
entire study on which the Main Group's sleeping grove is known (b.c.l. 0.00–0.03).

male was one of the lead members throughout the progression. Two females, Goldy and Null, and the most dominant male were among the lead members during parts of the progression.

By 1807, the group was near the western edge of the KB woods. They stopped and fed very rapidly, one might almost say desperately. Tension in the group remained high. About 1845, they ascended grove 2 b. The entire progression, from the time the baboons left the potential new grove, at 1719, until they ascended grove 2 b took 88 min and covered 2.5 mi., for an average of 1.7 mi./h.

Another grove, 11b, was actually adopted during our study[29]. So far as we know, it was used for the first time on the evening of May 12, 1964. The group was moving toward grove 11a, but at 1700 had reached an impasse in the vegetation, about half a mile north-east of the grove. Between them and the grove was a pan[30], fringed by dense, tall grass (*Sporobolus robustus*). The baboons were still in the same place half an hour later, when alarm barks were given by another group, about ¼ mi. to the southeast. (Sporadic barks continued until at least 1805.) By 1756, two baboons of the Main Group had moved to the nearest large grove of trees, 11b. Within 4 min, half the group had ascended the trees of 11b, and the rest were moving rapidly toward them. At 1820, we ascertained that another group was in the trees north of KH 4, in the area between groves 5 and 6. The sight of these baboons may have contributed to the Main Group's choice of the new grove.

The new grove, 11b, was subsequently used three times by the Main Group. The next time was 29 days later. At 1815, the group was moving across the same pan, but this time along a well-used route. They were moving toward grove 5. Suddenly, at 1824, there was a roar from a young adult male lion, seated on a ridge about 150 yards from the group and facing the group. The baboons gave their alarm bark, and ran to two small fever trees in the middle of the pan.

1828. The baboons were silent and motionless. So was the lion.

1829. About 10 members of the group left the trees and moved northward across bare soil.

[29] We are assuming that 11a and 11b are distinct. At the time of adoption, grove 11b had not been used on any of the 73 previous nights for which the sleeping grove of the Main Group is known (b. c. l. 0.000–0.055). This grove is known to have been used by another group at least once during our study.

[30] A natural basin or depression in the terrain that floods during the rainy season. Except for a border of *S. robustus* grass, pans in Amboseli are generally devoid of vegetation, perhaps as a result of accumulated salts.

1831. Seventeen of the baboons were on the ground, but they were sitting quietly, not progressing. An infant cackled, but there was no other vocalization.

1833. The entire group began moving northward, led by adult female Stocky, who was 15 ft. in front of the others. Half a minute later, two alarm barks were given. The baboons ran about 90 ft. farther north, then stopped and sat. They were completely in the open. The lion was still sitting quietly in the same place.

1834. About 5 members of the group moved southward again, i.e. toward grove 5. But the rest remained seated.

1836. About three-quarters of the group moved southward. Neither female Notch nor female Round Hips, both carrying neonates, showed any tendency to stay near adult males. But about 10 sec later, most of the group turned and moved northward again, in what looked like a more definite progression. An infant screeched—the only vocalization. The lion was still in the same place, unmoving.

Half a minute later, female Arch, with her infant clinging to her ventrum, moved southward, in the direction opposite from that taken by the group. The group stopped and watched her. About 40 ft. away, she sat down for perhaps one-half minute, then got up, moved another 10 ft. south, then sat down again. Virtually every baboon in the group was watching her. She sat there, looked back at the group, looked over toward the lion, and so forth. About 15 sec later, she got up and trotted toward and then through the body of the group. This action precipitated the rapid progression that followed, and the group moved directly to grove 11b.

By 1844, half the group had ascended the trees. The last three to ascend were two of the three immigrant males, Shag and Whitetip, and the partially paralyzed male, BBT.

It is interesting that female Arch, whose actions seemed quite decisive in resolving the uncertainty of the group, is the female who was carrying the youngest infant. Her son was just 7 days old at the time.

The other two uses of grove 11b were on July 6 and July 22. On the morning of the sixth, the baboons gave a false alarm while crossing this same pan, but they soon quieted down again. No other alarm reaction was noted that day. (We were away from the group from 1239 until 1635.) Utilization of grove 11b on the night of July 22, after almost adopting a new grove, has been described above.

One grove, No. 3 (fig. 19) was deserted after two members of the group were killed therein by a leopard (p. 182). Although this grove had been used as a sleeping grove on 13 out of the 57 preceding evenings for which the grove of the group is known, it was not used once in the subsequent 68 nights for which we have sleeping grove data. This difference is highly significant ($z = 4.16$, $P < 0.001$; FREUND, 1952, sec. 10.4).

4. Time of Descent

On 152 mornings, we estimated the median descent time of the Main
Group, that is, the time at which the median member of the group
descended to the ground. The distribution of these observed descent
times is shown at the top of figure 21. However, the distribution of
observed descent times is not an unbiased estimate of the true distribution: on some days the baboons had already descended when we
reached them, and thus the distribution of observed descents is biased
against early descents.

The simplest way to remove this bias is as follows. The probability that the baboons
will descend during any time interval can be estimated from the number of descents
observed in that time interval divided by the total number of days during which the
group was observed up, down, or descending during the interval. (Note that the total
includes days when we arrived before the onset of the interval, only to find that the
baboons had already descended.)

As we have shown elsewhere [WAGNER and ALTMANN, 1973], a more reliable
estimate of the probability $P(d^t_{t+\varDelta})$ that the descent time will occur during the time
period $(t, t+\varDelta)$ may be obtained from

$$P\left(d^t_{t+\varDelta}\right) = \frac{1}{N}\left[{}^0_tD^t_{t+\varDelta} + \sum_i {}_iD^i_{t+\varDelta} + \sum_i {}_iD^0_i \frac{{}^0_tD^t_i}{{}^0_tD^0_i} + \frac{{}^0_tD^t_{t+\varDelta} \cdot {}_{t+\varDelta}D^0_{t+\varDelta}}{{}^0_tD^0_{t+\varDelta}} \right.$$

$$\left. \cdot \sum_j \frac{{}_jD^0_j}{{}_{t+\varDelta}D^0_j} \right],$$

where ${}^a_bD^c_d$ denotes the number of days on which observations began between a and b
o'clock and the baboons descended between c and d o'clock. ${}_bD^c_d$ denotes the number of
days on which observations began at b and the baboons descended between c and d,
N is the total number of days in the sample, $t < i < t+\varDelta$, and $j > t+\varDelta$. Since each day's
sample is essentially a Bernoulli trial, binomial confidence limits may be placed on the
estimate for each time period.

The estimates that we obtained and their 95% confidence limits are
shown in figure 21, where the results are compared with the (biased)
results of plotting only observed descent times.

For the purpose of calculating confidence limits for each time period $(t, t+\varDelta)$, one must
associate with that period a sample size. That size is certainly no less than the number
$N_t = {}^0_tD^0_\omega$ of days on which we arrived by the beginning of the interval. In fact, since
some additional information is gained from the days on which we arrived after the
beginning of the interval, the 'true' sample size is no doubt larger than N_t—but probably
not as large as $N_t + {}_\omega^tD^0_\omega = N$. Those two values, N_t and N, determine an outer and an
inner pair of confidence limits for the period $(t, t+\varDelta)$, both of which are plotted in figure 21.

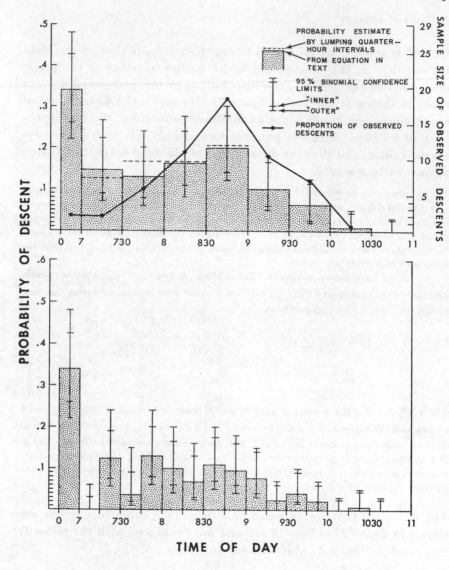

Fig. 21. Distribution of descent times, using half-hour intervals (upper graph) and quarter-hour intervals (lower graph). For explanation of estimation technique, see text.

5. Time of Ascent

The distribution of observed ascent times is shown in figure 22. Because of days on which we departed before the group had ascended, this distribution is biased against late ascents, but because of a considerable lack of independence between our departure time and the group ascent time, the earlier techniques are inadequate to remove the bias. At present, we do not know of any way to obtain an unbiased estimate from the available data. The mean of the observed ascents is 1822, with a standard deviation of 22 min.

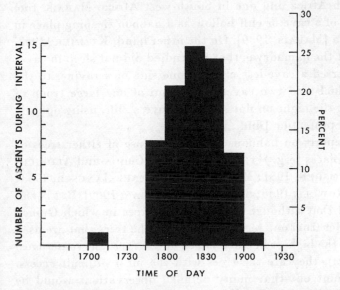

Fig. 22. Distribution of observed ascents of Main Group into sleeping trees.

6. Comparisons and Discussion

(a) Sleeping Sites

At night, the baboons that we observed in Amboseli slept only in trees. In the Serengeti plains, on the other hand, baboons were observed on a *kopje* (rocky outcropping). Except for several other kopjes nearby, we could see nothing but flat, treeless plains for miles around. In all likelihood, then, these baboons in the Serengeti slept on the kopjes. Since

the few trees on the kopjes seemed too small and too sparse to hold all
of the baboons, many if not all of them probably slept on the rocks
themselves, much like baboons that sleep on cliffs, such as the chacma
baboons in the Cape Province of South Africa [HALL, 1960], hamadryas
and anubis baboons in Ethiopia [KUMMER and KURT, 1963; CROOK
and ALDRICH-BLAKE, 1968], and perhaps also anubis baboons in the
arid Karamoja Province of Uganda [HALL, 1965 a]. In contrast, other
baboons that we observed in Serengeti, in the vicinity of Seronara, slept
in trees.

HALL [1963, p. 15] described the use of a cave by one group of chacma
baboons in South Africa and one in South-west Africa. MARAIS, too,
recorded the use of a cave or cliff hollow as a baboon sleeping place in
South-west Africa [MARAIS, 1939]. On the other hand, KUMMER [1968,
p. 157] notes that the hamadryas that he studied did not sleep in caves
and in fact preferred a cave-less cliff, on one side of a ravine, to the
opposite cliff, which had two caves. A portion of one large troop of
hamadryas spent the night on flat ground above a cliff, using opuntia
cactus as its only protection [ibid., p. 17].

All other reports on baboons indicate the use of either trees or
cliffs as sleeping places [e. g. HALL, 1963, 1965 b; CROOK and ALDRICH-
BLAKE, 1968; LUMSDEN, 1951; MAXIM and BUETTNER-JANUSCH, 1963;
DEVORE, 1962; ROWELL, 1966; STOLTZ and SAAYMAN, 1969]. BERT et al.
[1967a] observed that although the species of trees in which Guinea
baboons sleep differ from one region to another, the trees that are used
are, in each case, the highest in the area, or in one case (the *ronier* tree,
Boratius ethiopum), they are the trees with the most difficult access.
These authors point out that many detailed observations would be
necessary to detect any pattern in the allocation of branches within a
sleeping tree, but it seemed to them that females with infants took the
most peripheral positions, i. e. farthest from the trunk of the tree. In
Uganda, ROWELL [1966] noted that baboons used particularly tall trees
as resting places.

In many areas, there is no choice between trees and cliffs: only
one is present in suitable size. STOLTZ and SAAYMAN [1969] have studied
chacma baboons at Tshipise (23 mi. south of Messina in the northern
Transvaal, S. Africa), in an area where both are available. They write:

'The majority of the chacma baboon troops in the Messina district showed a
distinct preference for sleeping sites in the rocky krantzes [precipices] in the
sandstone ridges and kopjes; these were very similar to the steep cliffs used by

groups in the Cape and in South-West Africa [HALL, 1962]. A few troops, however, slept in trees and these utilized the tall trees on river banks *(Acacia albida,
Acacia xanthophloea, Ficus sycomorus, Croton megalobotrys)* in preference to
the large baobabs, *Adansonia digitata*, which were available throughout the
area. A small percentage of troops used krantzes and trees as sleeping sites when
both were available. Sleeping [cliffs in the study area]...had a typical smell and
were marked by accumulations of dung and the urine-stained, sticky surface of
the rocks[31]. In general, they were situated beneath overhanging rocks midway
between the upper and lower slopes of the cliffs.'

Hamadryas, when they have a choice, sleep on rocks rather than trees.
KUMMER [1968, p.161] writes:

'Strikingly, the permanently green, large (more than 15 m) trees of the gallery
forests were hardly used as sources of food and never as sleeping places. The
baboons only approached them when they went to drink and rest at the waterholes in the rivers, and then only a few animals would occasionally climb onto
limbs not higher than about four meters and pluck some fruits.'

The use of trees as sleeping sites is virtually universal among other
monkeys—probably a reflection of their basically arboreal habits. The
use of cliffs, caves and rocky outcroppings probably is relatively recent,
a trait that developed along with other adaptations to living on the
ground and moving into areas with progressively fewer trees. The presence of both tree- and rock-sleeping baboons in Serengeti indicates the
present-day adaptability of these animals.

    What is the significance of the fact that baboons sleep off the
ground, in trees, on cliff faces or on rocky outcroppings? We tend to
agree with DEVORE and HALL [1965], who say that 'the most general
statement it is possible to make about baboon sleeping sites is that they
seem to choose the safest places available to them'. BERT *et al.* [1967a]
concur, and suggest further that additional protection is gained through

---

    [31] HALL, too, commented on the accumulation of feces on the ledges and at the
base of sleeping cliffs [HALL, 1962a, p.200], and BERT *et al.* [1967a] have commented
on the accumulation of feces below sleeping trees. Of course, accumulated feces may
serve as fertilizer. When the same sleeping places are used for many years [HADDOW,
1952], an appreciable amount of material may be transported. Dispersion and germination of acacia seeds and those of other plants may be facilitated by being carried in
feces, and some hard, round seeds seem to be particularly adapted to passage through
the guts of animals without causing harm and without suffering damage themselves
[LAMPREY, 1967]. KUMMER [1968, p.17] has suggested that opuntia cactus is planted
in this way, and that locally, hamadryas may become independent of the scarce sleeping
rocks by planting such protective screens.

familiarity with the habitat that comes from repeated use of a favorite sleeping grove.

However, sleeping in trees is no guarantee of safety, as demonstrated by the killing of two members of the Main Group by a leopard (p. 182). This case tends to support our statement above, that the baboons are less subject to attack when several groups sleep near each other: the grove in which the Main Group was attacked by the leopard was one of their more isolated groves.

A few data on the number of sleeping sites per group have been published. During DeVore's study of olive baboons in Nairobi Park, 'one group, whose core area contained only one isolated grove of tall trees, slept in this spot on all but two evenings throughout 10 months of almost daily records, while groups that ranged along the [Athi] river might have as many as 14 sleeping sites' [DeVore and Hall, 1965, p. 32].

Hall's C group of chacma baboons in the Cape Reserve, South Africa, used three sleeping-cliffs. His S group used 3 cliffs in 1958/59, and the same 3, plus 2 others, during his 1960 observations. Frequencies of use are given and, for 3 groves, Hall lists the numbers of successive nights that the same grove was used [Hall, 1962a, p. 200].

Stoltz and Saayman [1969] give the frequencies of use of sleeping sites for two chacma groups in the Northern Transvaal, one of which used 5 sites, and the other, 6. Data on the order of use of sleeping sites are given for the former (troop W), whose most frequently used krantz accounted for 71.2 % (94/132) of the nights in the sample. For that favorite krantz, the chance that the group would return if they had slept there the previous night was about 0.78 (= 32/41). Thus, if this krantz exerts any residual attraction of the sort shown by Sleepy Hollow in our study, it is not obvious from the available data.

Buxton [1951] presented evidence that baboons and other species of monkeys in the Mongiro Forest of Uganda may return to certain sleeping trees year after year.

Kummer [1968] does not indicate the frequency with which various sleeping rocks were used by particular one-male units of hamadryas baboons, but he does point out that a rock that is used by many baboons on one night may be vacant the next. He also suggests that abundance of available rocks may affect troop size. 'Troops are large (up to 750 individuals) in areas where rocks are scarce, and small (as few as 12 individuals) where rocks are abundant' [Kummer, 1968, p. 148].

(b) Sleeping Clusters and Subgroups

In Amboseli, the concentration of several groups of baboons into one area at night seems to reflect not only a concentration of suitable trees, but also an element of sociability between groups. Sleepy Hollow, the Main Group's most frequently used sleeping place, did not seem to us to be any better or worse than their other sleeping places. What it did have to offer was proximity to other baboon groups. We do not known the function of this evening convergence of groups. Perhaps any predator in the area is more likely to be seen by the baboons.

This night-time concentration of baboons is quite unlike what has been reported for baboons in Nairobi Park [DeVore, 1962]. It tends towards the pattern found in hamadryas baboons that has been reported by Kummer and Kurt [1963] and Kummer [1968], in which many one-male parties converge each evening on an area of cliffs used for sleeping. Similarly, Stoltz and Saayman [1969] write: 'A sandstone ridge, six miles in length, intersected the central part of the study area.... Whereas a large number of potential sleeping places were available, five troops... utilized sites concentrated within one mile of the ridge.' Aggregations of Guinea baboon groups at groves of gallery forest have been described by Bert et al. [1967a], but in such areas, the aggregations may merely represent the concentration of suitable trees.

We have indicated that in Amboseli, groups of baboons aggregate in the evening in the vicinity of their sleeping groves, which tend to be clustered, and that our main study group occasionally split into two subgroups in the evening, each subgroup sleeping in a different grove of trees. A similar combination of clustering and subgrouping was described by Zuckerman [1932, p.199], based on his observations of chacma baboons in South Africa. He wrote:

'A large pack of about one hundred animals may occupy more than one rocky cliff or krantz when it settles for the night. During the day the baboons may at times have been widely scattered, but towards evening they reunite and huddle in small parties in crevices that nearly always open onto a slope.'

Lumsden [1951] carried out a detailed study of the night-resting habits of several species of monkeys, including olive baboons, in the Mongiro Forest of western Uganda. The two favorite sleeping trees of the local baboons were both *Mitragyna stipulosa*, of about 'average' (25 m.)

height. The crowns of both communicated in many directions with the crowns of other trees of the same species and of *Elaeis guineensis*. The 5 favorite sleeping trees of the baboons 'accounted for 72 % of all *Papio doguera tessellatus* bands observed' [ibid., p. 21]. About three years later, a restudy of the same forest revealed that at least 4 of these 5 favorite baboon sleeping trees were still being used by them [BUXTON, 1951]. In their choice of sleeping places, the baboons favored

> 'particularly the forest fringe and the open ground closely adjacent to it. In the case of their two most frequented trees...little indication was evident on the matter of preference of some trees over others; *M. stipulosa* is overwhelmingly the commonest tree in the swamp forest and all specimens are remarkably uniform in size and formation.... Yet [the favorite tree] was also highly frequented by *Cercocebus albigena johnstoni* [black mangabey] so that it appears that it possessed characteristics which rendered it a popular sleeping tree for more than one species' [ibid., p. 22].

LUMSDEN [1951] reported the distribution of sleeping band sizes in eight species or subspecies of primates in the Mongiro Forest. He got a mean band size of 4.66 for olive baboons, *Papio doguera*, by discarding data on solitary individuals; if one includes these solitary animals, mean size for these baboon sleeping bands was 4.43. Yet, 'in the daytime *P. d. tessellatus* was frequently seen in a band numbering about 40 individuals but also in smaller ones of only 16 to 20. The sleeping parties apparently separate from the band in succession at about sunset' [ibid., p. 29].

HALL [1962 a] has summarized his observations on the groupings of savannah baboons at their sleeping places:

> 'The cliffs most frequently used by S group in the Cape consisted of steep, usually vertical, rock-faces rising more or less sheer from the sea.... Reliable observation was here almost impossible, but many observations of S at Matrooskop and C at Paulsberg and Kommetjieberg have been made from fairly close range before dawn, and after dusk when there has been sufficient moon for binoculars to be used. Observations have also been made in South-West Africa from hides directly opposite the sleeping-cliffs. The baboons in view have been observed mostly huddled in twos and threes, with a few single animals, and usually in the sitting posture, on a ledge vertical to the rock-face or crevice. One baboon was observed lying on his side, on a ledge, dog-like, with tail dangling down.... [LUMSDEN] suggests that...sleeping parties separate from the larger band in succession at about sunset. There is, however, no way of determining whether such sub-grouping has any social significance, or whether it is simply a convenient adaptation to the physical requirements of tree-space or rock-formation, allowing for reasonable warmth, shelter and protection from predators.

The possibility remains, as a result of our own observations, that all or most animals in a group sometimes pass the night close together. Thus, in South-West Africa, we observed from a hide the whole of a group of 23 animals emerging at dawn from a single shallow cave in the cliff-face. Watching C group from below their Paulsberg sleeping-place, we have several times seen only a few animals emerge before dawn, individually or in twos, from rock-crevices, the majority of the group appearing *en masse* later from behind the shelter of thick bush at the base of the cliffs.'

(c) Descent and Ascent

Several authors have commented on the time of day at which baboons descend from or ascend into their sleeping trees or cliffs, and in some cases the results are strikingly different from what we observed in Amboseli. For example, STOLTZ and SAAYMAN [1969] write:

'Baboons approached sleeping sites in the late afternoon and by dusk were occupying the cliffs. A troop might start the ascent as early as 1530 h and progress slowly in scattered groups to gather at the sleeping site as darkness fell.... Troops started to leave the sleeping site at first light when animals were able to investigate the surroundings. Progression started earlier than usual when conditions of bright moonlight prevailed in the early morning. At daybreak (0450 h in summer, 0600 h in winter) mature males and juveniles were generally the first to show signs of activity.... In winter a troop might move only a short distance at first light to take shelter from cold winds; large males, in particular, remained sitting for long periods with their backs to the sun in the vicinity of the sleeping site.'

In contrast, ROWELL [1966], in describing the olive baboons that she observed in a Uganda forest wrote that,

'baboons are not early risers, and they rarely left the sleeping trees before 8:00 a.m. (dawn 6:00–6:30 a.m.). They usually fed in the sleeping tree and neighboring trees until midmorning, feeding gradually giving way to play and grooming and finally dozing. After this a move would be made to another part of the forest, or out to feed in the open grass or one of the areas of bush or isolated trees....'

According to LUMSDEN's report [1951], the olive baboons of the Mongiro Forest, Uganda, entered the sleeping trees during the hour before sunset (and perhaps at other times); they become inactive by half an hour after sunset. The next morning, almost all descended before dawn.

KUMMER [1968] describes the departure of hamadryas baboons from their sleeping rocks as follows:

'At about sunrise the parties leave their ledges in the sleeping rocks and go to the open waiting areas above the face of the cliff to sit in the sun.... The time spent in waiting areas can vary from a few minutes to as much as three hours. During this time the adult animals will doze, heads sunk on their chests, or will devote themselves to social grooming within the one-male unit. Chasing between adult males and copulation primarily occur during this morning rest period, while young animals form play groups. At the same time the troop is preparing for its departure: The shifts of the individual one-male units become more and more frequent' [p. 12].

'In the Erer area, the time of departure in the morning varied as much as 1 h and 50 min in the same troop from one day to the next, and between 06.00 h and 10.00 h official time throughout the months from February through October. There is a positive rank order correlation of the morning schedule with the monthly average time of astronomical sunrise. The last animals left their sleeping ledge 28 to 38 min after sunrise (R = 0.94, P<0.05), and the troop's departure from the waiting area followed the time of sunrise by 55 to 120 min in the monthly averages (R = 0.83, not significant). Not included in this calculation are the months of the long rains, July and August. During this time, on rainy as well as on sunny mornings, the animals waited twice as long before they left the cliff and before they departed from the area' [p. 167].

The seasonal shift observed by KUMMER, as well as the very early departures of baboons in the Transvaal [STOLTZ and SAAYMAN, loc. cit.] and the late departures of the baboons in the verdant Ishasha River valley [ROWELL, loc. cit.] suggest that the time of descent or departure from sleeping sites may be correlated with the total productivity of the area, and thus with the amount of time that the baboons must devote to foraging in order to sustain themselves. HALL [1962 a, p. 201] concurs and offers an additional suggestion: 'The much earlier leaving times of the South-West Africa groups could be due to the much colder early-morning temperatures or to the need to make much longer day-ranges in order to obtain a sufficiency of food.' HALL [ibid.] also noted a tendency for the baboons to depart 'later in the summer months, when the amount of day-time is greater than in winter', but the difference in means was not significant. (HALL's comparisons are based on minutes before or after sunrise, not time of day.)

According to STOLTZ and SAAYMAN [1969], 'troops [of chacma baboons in the Transvaal] did not attempt to occupy sleeping sites until they had scanned the cliffs at length from beneath. The gradual approach of the baboons to the sleeping sites indicated a marked degree of caution. In general, a single dominant male preceded the troop by approximately 50 yards. The troop followed slowly and vocalizations were conspicuously absent. Similar behavior was not observed at sites 1 and 2, where troop W was unlikely to encounter other troops'.

STOLTZ and SAAYMAN also indicate that when approaching sleeping trees in gallery forest, 'their behavior lacked the silence, slowness and caution which characterized the approach of troops to adjacent sleeping cliffs.'

According to BERT et al. [1967a], Guinea baboons go to their sleeping trees at 1830 to 1915 and usually ascend immediately. In three cases, the troop was preceded some 20 to 30 min earlier by a small subgroup composed of an adult male and several juveniles, an observation that is similar to that of STOLTZ and SAAYMAN [above].

(d) Vantage Points

Trees, cliffs and outcroppings also serve as vantage points for baboons. From them, they have a view of much of the surrounding countryside. No doubt this is advantageous for detecting predators as well as for keeping track of other groups, for finding water, and so forth. Repeated use of trees as vantage points by an isolated male was described in chapter III.

The significance of trees and cliffs as vantage points has also been commented on by several authors. Describing anubis baboons in Ethiopia, CROOK and ALDRICH-BLAKE [1968, pp. 214–215] write that these animals 'commonly climbed into trees—the large adults to sit ponderously on the larger branches.... Motionless adults in trees had a good view through the forest and would often sit there barking occasionally at some distant disturbance.'

HALL [1960] described vigilance behavior and visual scanning in chacma baboons, and reviewed the earlier literature on 'sentinal' behavior. Rocks or trees are sometimes utilized for this purpose. These observations have been corroborated and extended by MAXIM and BUETTNER-JANUSCH [1963], who describe a triangular formation of vigilance in what are probably yellow baboons, living in an area covered by a heavy growth of scrub bush. According to these authors, this particular pattern was assumed whenever, and only when, the baboons approached an area of known danger or were surprised by the observer. Family units always remained within the triangle.

STOLTZ and SAAYMAN [1969] write: 'The large males climbed trees and scanned the area whenever vocalizations such as barking, squealing or the copulatory calls of the females indicated that another troop was in the vicinity of the drinking site.'

V. GROUP MOVEMENTS

1. Diurnal Cycle of Activities

After descending from their sleeping trees, but before moving to the foraging grounds, the baboons sometimes stayed in the vicinity of the trees for a while, going through a period of intensified social activity. On other days, they moved directly away from the sleeping grove to a foraging area.

If this foraging area was in open grassland, there was a clear-cut difference between the formation of the group while moving to the feeding area and their formation while feeding. During the former, they moved in a long *file* (not necessarily single), whereas when foraging in open grassland they usually moved slowly forward in a *rank*, i.e. with the long axis of the group roughly perpendicular to the line of progression [c.f. ROWELL, 1966; STOLTZ and SAAYMAN, 1969]. Since the front of the file reached the foraging area before the rear, and spread out earlier, the group went through a triangular transition stage, something like that described by MAXIM and BUETTNER-JANUSCH [1963].

Late in the afternoon, the group would return to its file formation and move toward the groves of sleeping trees. On many afternoons, there was a period of intensified social activity on the ground, in the vicinity of the sleeping trees.

A crude quantitative picture of the diurnal cycle of social behavior was obtained by tabulating the number of index entries for each pattern of social behavior during each hour of the day, lumping these behaviors into major categories, then plotting, for each category, the number of index entries per hour of close observation on the Main Group (fig. 23). These values may be taken as rough estimates of the number of social acts per hour in this group.

In figure 23, the large morning and evening peaks in social interactions conform with our impressions from the field. The smaller peak at 1300–1500 was unexpected, however. It occurred at a period of day in which our sampling (field time) was comparatively low, and may, therefore, be an artifact.

We tested the null hypothesis that the number of social interactions per hour of observation during the two-hour period 1300–1500
was not greater than that for the period composed of the immediately
preceding hour and the immediately following hour, i.e. that the 'peak'
was not significantly higher than its adjacent 'valleys'. The testing was
done as follows.

Let t_1 and t_2 be the amount of time (i.e. number of hours of observation) in
which the n_1 and n_2 behavioral events in the 'peak' and 'valley' periods (respectively)
occurred, and assume that the behaviors during each time period may be approximated
by Poisson processes with rates λ_1 and λ_2, respectively. Our null hypothesis is simply
that $\lambda_1/\lambda_2 = 1$, to be tested against the one-sided alternative that $\lambda_1/\lambda_2 > 1$. This can be
done by calculating the variance ratio $(n_1 t_2)/(n_2 t_1)$ and referring to a table of the
distribution of F, with $(2n_1, 2n_2)$ degrees of freedom [see COX and LEWIS, 1966, p. 229][32].
Our data give a variance ratio of 1.3599, which, with (1557, 1517) degrees of freedom, is
significant at the 0.05 level; in other words, the activity peak between 1300 and 1500 is
statistically significant.

While this observed peak in social activity between 1300 and 1500 is
apparently not a small-sample anomaly, it may represent the result of
biased sampling. We sometimes terminated observations at about mid-
day, at which time the baboons were often foraging, widely dispersed,
and seldom interacting. But doubtless, we sometimes remained longer
if 'something was happening', e.g. if the baboons were particularly
interactive. If so, an exaggeration in the apparent amount of social
activity during the subsequent few hours would result. For the field
worker, the research implication of this bias is straightforward: the
time at which each period of observation begins and ends must be
independent of what the animals do during that period.

Much of the time between the morning and evening 'social hours'
is taken up with feeding and foraging (feeding 'on the move'). On 104
occasions, we recorded the predominant activity in the Main Group.
In general, this was done at the end of systematic time samples on an
individual's behavior, but it was also done on other occasions. The results, which are given in table XIV, indicate that feeding or foraging is
usually the predominant activity in the group during the middle hours
of the day.

---

[32] This same variance ratio technique was used beforehand, in an attempt to
justify the pooling of 'valley' data from 1200–1300 with those from 1500–1600. The
calculated variance ratio was 1.0772 with (1870, 1152) degrees of freedom. We have not
been able to find a table of F sufficiently extensive to evaluate this case.

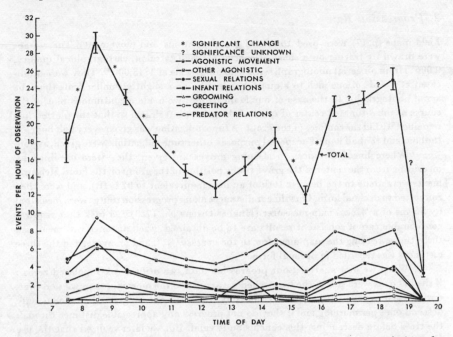

Fig. 23. Diurnal cycle of social behaviors, compiled by an approximate technique described in text. Points on the upper line (TOTAL) are the sum of the points below them. Vertical bars represent 95 % confidence limits [Cox and Lewis, 1966, p. 31, equation 8]. Significant changes in rate (indicated by *) were tested statistically by a method described in the text.

Table XIV. Predominant activity of baboons in the Main Group, sampled on 104 occasions

Time of day	Number of cases	
	Feed/Forage	Other
0800–0900 h	4 (44 %)	5
0900–1000 h	11 (79 %)	3
1000–1100 h	14 (100 %)	0
1100–1200 h	10 (83 %)	2
1200–1300 h	6 (100 %)	0
1300–1400 h	8 (100 %)	0
1400–1500 h	4 (80 %)	1
1500–1600 h	8 (89 %)	1
1600–1700 h	6 (75 %)	2
1700–1800 h	9 (69 %)	4
1800–1900 h	3 (50 %)	3

2. Progression Rate

Field maps (p. 17) were used to record group positions and movements. These maps were drawn by tracing on a sheet of special plastic (Astralon, cartographical quality, 0.006″) from an aerial photograph with a nominal scale of 1:15,000[33]. They were duplicated at a scale of one inch to a quarter mile, i.e., on a slightly smaller scale than the aerial photographs. In the case of our Main Group, we made a continuous plot of the course of the estimated center of mass of the group, but drawn so that the plotted line remained within the confines of the group. A time indication was given every half hour, on the hour and the half hour. For special purposes, other time indications were given, as well.

Where linear distances to sleeping groves are given, they were obtained by measuring from the center of the grove to the position of the group on the hour. Measurements were made to the nearest 16th of an inch (equivalent to 82 ½ ft.), and were then converted to decimal form. Curvilinear distances along progression routes were measured by means of a 'Hoco' map-measurer (Hughes-Owens No. 1723)[34], a task that requires considerable care if consistent results are to be obtained. Each distance was measured three times, reading the map measurer to the nearest $1/16''$, and an average of the three distances was recorded in decimal form.

When we first analyzed our progression data, we utilized 'extrapolated zeros': if the baboons were still in the trees when we arrived in the morning, then we said that, during each previous time interval that day, they did not progress. If baboons only descend once per morning and if they do not progress any appreciable distance through the trees before descending, this conclusion is valid. But we later realized that the use of such data is biased against those days on which the group descended before our arrival, and thus utilizing such extrapolated-zero distances would lead one to underestimate the mean distances progressed. All such 'data' were eliminated.

Bias of another sort is to be found in table XXII, XXVI, XXIX, and XXX, and results from the fact that we inadvertantly used distances from some partial intervals, i.e. from half-hour periods in which we were not continuously present. This bias results in an underestimate of true mean distances progressed and would be most marked during the morning hours because our observations generally began part way through a time period. A technique somewhat like that utilized for estimating the distribution of descent time (p. 81) probably can be devised to make use of actual data obtained when the observer arrives or leaves part way through an interval. In lieu of this, the field worker should consider beginning and terminating observations at the boundaries of his standard time intervals, whenever practical.

Since the movement during foraging was typically much slower than the preceding movement to the foraging areas, the progression rate of the group tended to have a maximum each morning, generally between 1000 and 1030. There was, however, much variability in the diurnal cycle of progressions. Figure 24 shows, for each half hour of the day, an

---

[33] No ground controls were available for these aerial photographs. The scale indicated here was provided by the cartographic section of the Survey of Kenya office.

[34] A map measurer with a smaller tracing wheel would be preferable.

estimate of the mean, standard error of mean, standard deviation and range of distances traveled by the Main Group.

Figure 24 includes data from periods during which the group was still in the trees. Once the baboons were on the ground, however, the rate of progression was as shown in figure 25. (Because no descents were observed after 1027, this distribution would be identical with that shown in figure 24 for subsequent time intervals.)

It may be that the distance traversed by the baboons during any period of the day depends more upon the time elapsed since the group descended than upon the actual time of day. Figure 26 shows the distri-

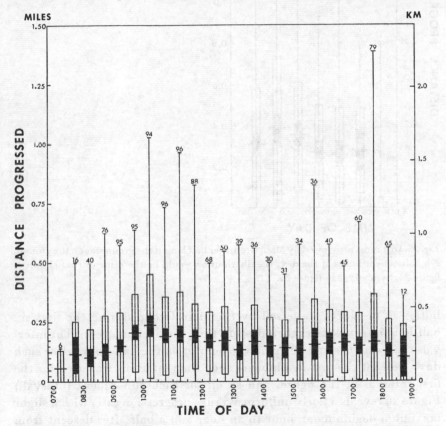

Fig. 24. Distance progressed by the Main Group during each half hour of the day. For each half hour, the graph gives the mean (horizontal line), standard error of the mean (black bar), standard deviation (clear bar), and range (vertical line) of distances progressed; sample sizes are indicated above each interval. Maximum likelihood estimates are given for means. Maximum probability estimates are given for standard deviations

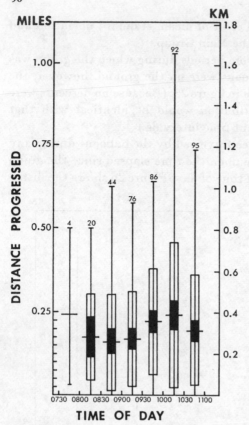

Fig. 25. Distances progressed by Main Group per half hour if the group was on the ground. For subsequent hours of the day, the distribution would be the same as that shown in figure 24. Symbols as in figure 24.

bution of distances traversed by the Main Group during the first six half-hour intervals after their descent from the sleeping trees. (The intervals that were used all begin on the hour or on the half hour; after each day's descent the first available of such time intervals was used for the first interval in the figure, and so on for each subsequent interval.) Figure 26 reveals a fairly uniform average progression rate, with a slight maximum beginning an hour to an hour and a half after descent from the trees.

The distribution of progression rates during any time period is far from symmetric; rather, most progressions are slow, and very rapid progressions are relatively rare. This fact may be appreciated from

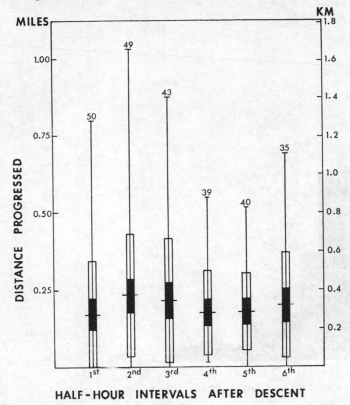

Fig. 26. Distance progressed by Main Group per half hour after descent. Symbols as in figure 24.

figure 27, which shows the distribution of distances progressed per half hour in the periods 1030–1100, and 1100–1130. (This hour was chosen because it gives a reasonably large sample, 192.)

3. Day-Journeys

An indication of the total distance traversed by the Main Group on an average day can be obtained by adding the mean distances traveled during all half-hour intervals. These distances can be read from the end-points of the cumulative graphs given in figure 28. For the whole study—that is, disregarding seasonal variations—the cumulative mean total distance traversed by the Main Group is 3.67 mi. per day. An

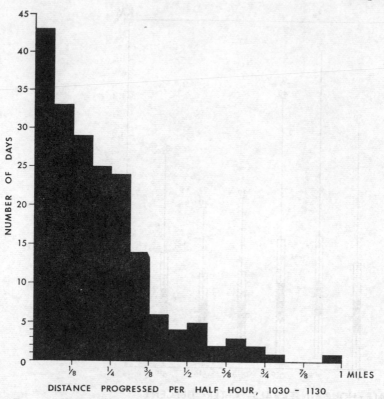

Fig. 27. Distribution of distance progressed by Main Group per half hour between 1030 and 1130.

alternative method is to utilize data from just those days for which we have complete day-journey records. These give a mean distance of 3.42 mi. ($S.E. = 0.24$ mi., $S.D. = 0.84$ mi., $N = 12$).

Seasonal variations were slight except for apparently longer progressions during the January-February period. The full-day cumulative total for the January-February period, which is 34 % higher than the overall, nonseasonal total, includes 5 fictitious data points (see caption of figure 28), and may therefore be misleading. However, it should be noted that for this season the cumulative total through 1300 h—the latest time of day that is not based on any fictitious points—is already 27 % higher than the nonseasonal total at that time of day. At present, we have no explanation for the anomalous total during this period, nor for the relative uniformity during the rest of the year.

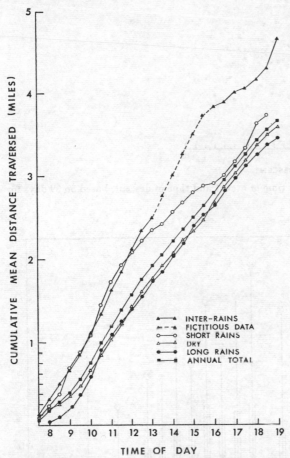

Fig. 28. Seasonal variations in the total distance progressed by the Main Group. For each time of day, the graph indicates the average distance traversed by the group before that time. (For the inter-rain period, no data are available for the period 1330–1600. Points plotted for this period are fictitious; they are based on the mean of the preceding four points on that line.)

KUMMER [1968, p. 167] wrote of hamadryas baboons that 'the duration of the daily march tends to be constant. When a troop departs before 08.00 h it will significantly more often be back at the rock before 17.00 h than when it departs later (P = 0.03).' For the Main Group of yellow baboons in our study, the time of both ascent and descent of the sleeping trees is known for 37 days. The distribution of ascent vs. descent times is shown in figure 29. It does not reveal the kind of trend that KUMMER describes.

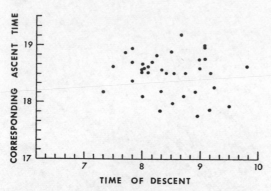

Fig. 29. Relationship between time of ascent and time of descent, based on 39 days for which both times are known.

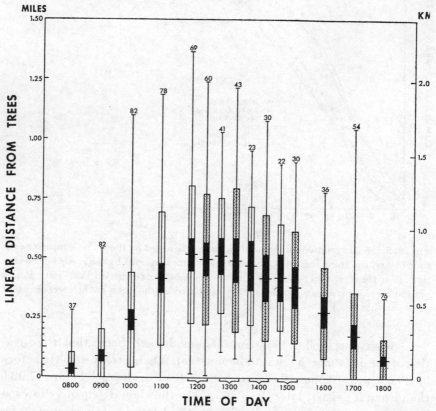

Fig. 30. Linear distance of Main Group from sleeping trees. For each time of day, the graph shows the distance from their sleeping grove of the last night (clear bar) or the distance to the sleeping grove of the coming night (dotted bar). Symbols as in figure 24.

4. Proximity to Sleeping Trees

Once the baboons had begun their morning progression, they showed no further attraction to the proximity of their sleeping grove of the last night; to the contrary, they generally moved progressively farther away, reaching a maximum linear distance from the sleeping groves about noon, or about 5 h after descent. Each afternoon, in turn, they got progressively closer to the coming night's sleeping grove. These spatial relations are depicted in figures 30 and 31.

A résumé of data on seasonal variations in the distance of the Main Group from the previous night's grove and distance to the coming night's grove is given in table XV. While some of the sample sizes in the table are small, an interesting comparison can be made between the data for the long rains (March–April) and for the dry season (May–Octo-

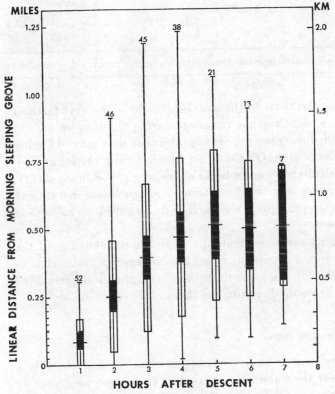

Fig. 31. Linear distance of Main Group from sleeping grove of last night at hour intervals after descent from the grove. Symbols as in figure 24.

Table XV. Seasonal changes in linear distance of Main Group from their morning sleeping gr or to evening sleeping grove

| | Jan-Feb (inter-rains) | | | | | | Mar-Apr (long rains) | | | | | |
| | Distance from grove | | | Distance to grove | | | Distance from grove | | | Distance to grove | | |
	n	x̄	s.d.	n	x̄	s.d.	n	x̄	s.d.	n	x̄	s.d.
0800	10	0.38	0.66				13	0.05	0.11			
0900	8	1.13	1.10				26	0.50	0.64			
1000	8	3.35	2.52				24	1.60	1.33			
1100	8	3.83	2.84				22	3.59	2.43			
1200	5	4.76	2.78	4	3.74	2.30	21	3.98	2.35	17	4.78	2.01
1300	2	4.44	2.47	3	5.48	3.49	14	3.68	2.17	13	4.37	2.25
1400	0	–	–	–	–	–	9	2.93	2.00	12	3.62	2.13
1500	0	–	–	–	–	–	7	2.39	2.19	9	2.95	2.24
1600				–	–	–				9	2.20	1.94
1700				2	0.37	0.18				14	1.42	1.26
1800				8	0.71	0.60				25	0.68	0.83

Distances are given in one-quarter mile units. n = sample size; $\bar{x}$ = mean; s. d. = standard deviati

ber). These data suggest that, on the average, the baboons were farthest from their previous night's grove by noon during the long rains, but not until about 1400 during the dry season, and that they moved farther away during the dry season. On the other hand, during the long rains, the group approached the coming night's sleeping grove more slowly, in that at almost every hour of the afternoon, their linear distance to the coming night's sleeping grove was farther during the long rains than during the dry season.

The morning progressions generally took the baboons out of the forests and onto the open grassland, where they foraged. Social activities became minimal (fig. 23). On some days, however, the group remained within the forest the entire day, and fed there.

5. Change in Progression Rate

Baboon group progressions have a strong inertia, in that there is a marked tendency for the baboons to travel as far during any hour as during the next or the preceding hour. This tendency is clearly indicated in figure 32. Much of the inertia is accounted for by sustained slow (or

Table XV. (continued)

May-Oct (dry)						Nov-Dec (short rains)					
Distance from grove			Distance to grove			Distance from grove			Distance to grove		
n	$\bar{x}$	s.d.	n	$\bar{x}$	s.d.	n	$\bar{x}$	s.d.	n	$\bar{x}$	s.d.
9	0.02	0.06				5	1.02	0.92			
41	0.48	0.52				7	2.26	1.43			
43	1.53	1.17				7	3.67	1.56			
42	2.76	1.81				6	5.44	2.32			
36	3.83	1.97	35	3.67	2.32	7	5.61	3.36	4	3.37	1.56
22	4.04	1.72	23	3.60	2.53	3	6.10	1.86	4	3.33	1.71
13	4.30	1.93	15	3.18	2.28	1	4.38	0	3	3.04	1.75
14	3.85	1.48	18	3.21	1.80	1	3.75	0	3	2.40	1.47
			24	2.19	1.47				3	2.19	1.14
			34	1.49	1.64				4	1.25	0.79
			38	0.56	0.69				5	0.40	0.37

zero) progressions: 49 % of all hour-to-hour transitions were cases in which the Main Group traveled less than $^3/_8$ mi. during each of two consecutive half-hour intervals (table XVI).

Figure 32 and the other data given below on change in progression rate were compiled in the following manner. For each hour of the day, a matrix of transition frequencies was drawn up; the rows indicate the rate of progression during the hour in question, and the columns indicate the rate of progression during the next hour of the day. Table XVI, which was used to make figure 32, is the sum of these hourly matrices. Thus, these tabulations are based on all pairs of consecutive hours in which the length of progression of the Main Group is known during both hours. The rows and columns were in $^1/_8$ mi. increments. Those cells that represent the same change in progression rate will all fall along the main diagonal of the matrix (no change) or along one of the diagonals that is parallel to it, with accelerations on one side of the main diagonal, decelerations on the other.

The data that are shown in figure 32 (dots and solid line) are compared with two null hypotheses, the random hypothesis (dashed line) and the independence hypothesis (dotted line). The random hypothesis is that all progression rates from 0 to $2\frac{1}{2}$ m.p.h. are equally likely and the progression rate is independent from hour to hour, i.e. that all of the cells in the transition matrix of table XVI are equiprobable. The in-

Table XVI. Change in progression rate of Main Group

	Distance progressed during next hour (miles)														
Distance progressed during first hour (miles)	$0-\frac{1}{8}$	$\frac{1}{8}-\frac{1}{4}$	$\frac{1}{4}-\frac{3}{8}$	$\frac{3}{8}-\frac{1}{2}$	$\frac{1}{2}-\frac{5}{8}$	$\frac{5}{8}-\frac{3}{4}$	$\frac{3}{4}-\frac{7}{8}$	$\frac{7}{8}-1$	$1-1\frac{1}{8}$	$1\frac{1}{8}-1\frac{1}{4}$	$1\frac{1}{4}-1\frac{3}{8}$	$1\frac{3}{8}-1\frac{1}{2}$	$1\frac{1}{2}-1\frac{5}{8}$	$1\frac{5}{8}-1\frac{3}{4}$	$1\frac{3}{4}-1\frac{7}{8}$
$0-\frac{1}{8}$	23	19	16	8	8	3	0	1	2	0	0	0	0	0	0
$\frac{1}{8}-\frac{1}{4}$	16	19	9	9	10	2	2	0	1	2	0	0	0	0	0
$\frac{1}{4}-\frac{3}{8}$	14	19	18	13	7	5	0	0	2	0	0	0	0	0	0
$\frac{3}{8}-\frac{1}{2}$	8	5	18	18	3	8	3	0	1	0	1	0	0	0	0
$\frac{1}{2}-\frac{5}{8}$	1	8	7	9	2	3	3	1	0	0	0	0	0	0	1
$\frac{5}{8}-\frac{3}{4}$	0	1	1	2	3	3	1	0	2	0	0	0	0	0	0
$\frac{3}{4}-\frac{7}{8}$	0	2	1	2	1	2	1	0	1	0	0	0	0	0	0
$\frac{7}{8}-1$	0	0	1	1	0	0	0	1	0	0	0	1	0	0	0
$1-1\frac{1}{8}$	0	2	0	2	2	1	0	0	0	0	0	0	0	0	0
$1\frac{1}{8}-1\frac{1}{4}$	0	0	1	0	0	1	0	0	0	0	0	0	0	0	0
$1\frac{1}{4}-1\frac{3}{8}$	0	0	0	0	0	0	1	0	0	0	0	0	0	0	0
$1\frac{3}{8}-1\frac{1}{2}$	0	1	0	1	0	0	0	0	0	0	0	0	0	0	0
$1\frac{1}{2}-1\frac{5}{8}$	0	0	0	0	0	0	0	0	0	0	0	0	0	0	0
$1\frac{5}{8}-1\frac{3}{4}$	0	0	0	0	0	0	0	0	0	0	0	0	0	0	0
$1\frac{3}{4}-1\frac{7}{8}$	0	0	1	0	0	0	1	0	0	0	0	0	0	0	0

The table entries indicate the frequency with which the group was observed to progress the distance indicated for the rows during one hour and for the columns during the second. Data that fell on the border between two intervals were placed in the upper interval.

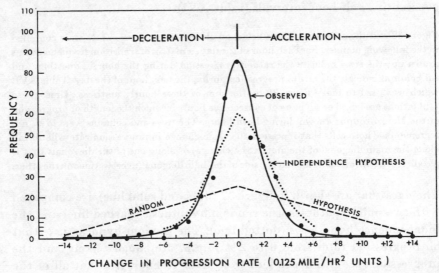

Fig. 32. Distribution of change in progression rates of the Main Group. See text for detail.

dependence hypothesis is the one that is routinely used in the analysis of contingency tables, namely, that the probability that an observation will fall in any row can be estimated from the contribution of that row to the grand total, and similarly for the probability of being in any column, and that row and column probabilities are independent of each other.

Neither hypothesis fits the data. For the independence hypothesis, which has the closer fit, $\chi^2 = 256.478$ (rows calculated with columns 12–14 pooled, 144 d.f.), which is significant beyond the 0.0005 level. As may readily be seen in the figure, almost all of the deviation from these two hypotheses is accounted for by the high frequency of small changes in progression rate. Thus, the inertia in the progressions of the group cannot be accounted for by either model.

Figure 32 suggests that changes in progression rate may have a normal distribution. However, the observed distribution deviates significantly from a normal distribution with the same mean and variance: $P(\chi^2 = 307.184; 20 \text{ d.f.}) < 0.001$. In particular, the tails of the observed distribution are too low. Similarly, despite the appearance of these data

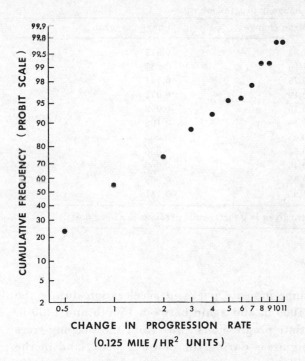

Fig. 33. Probit analysis of change in progression rate. See text for details.

when 'folded' on zero and plotted on logarithmic probit paper (fig. 33), the observed data deviate significantly from a log-normal distribution ($P < 0.001$, $\chi^2 = 30.59$, 9 d.f. with last two cells pooled). The nature of the underlying distribution is at present unknown.

The rate of acceleration changed during the day. The diurnal shift in acceleration may be obtained in two ways, first, from mean changes in progression rate at each time of day (i.e. using the hourly matrices upon which table XVI is based) and second, from first differences in mean progression rate, at each hour, as given in figure 24. That is, we can use the mean change in progression rate, or the change in mean progression rate. The results are given in table XVII. The two methods give results that agree in sign for all two-hour periods except those beginning at 1200 and 1300 h. For these periods, sample sizes are small and the differences in rate are slight.

Table XVII. Diurnal shift in acceleration of Main Group

Beginning of two-hour period	Change in progression rate	
	Mean change	Change in mean
0700	2.33	0.077
0800	1.10	0.237
0900	0.75	0.141
1000	−0.19	−0.071
1100	−0.16	−0.098
1200	0.19	−0.102
1300	0.15	−0.008
1400	0.04	0.006
1500	0.48	0.069
1600	−0.55	−0.009
1700	−1.83	−0.141

Changes in progression rate are given in quarter-mile per hour per hour units.

6. Maxima and Minima

The data on mean change indicate peaks in acceleration during the period from 0800 h to 1000 h, and again between 1500 h and 1700 h. The group tends to initiate progressions away from the sleeping trees and toward the foraging areas during the morning peak, and in the reverse direction in the afternoon.

Despite the 'inertia' in group progressions, movement of the group nearly ceased for a while on most days, typically in the afternoon, and often when the baboons were in the vicinity of trees. During these 'rest periods', some members of the group would often climb into the acacias and feed, or just sit quietly in the trees looking about ('sentry' behavior), while others would remain on the ground, resting, playing, involved in other social behavior, or continuing to forage.

The magnitude of such progression minima, during the period 1100–1600, and their influence on the rate of progression during preceding and succeeding time periods, are shown in figure 34, based on a sample of 31 days for which the distance that the Main Group progressed is known for all half-hour intervals between 1100 and 1600. For each of these days, we determined the half-hour interval during which the

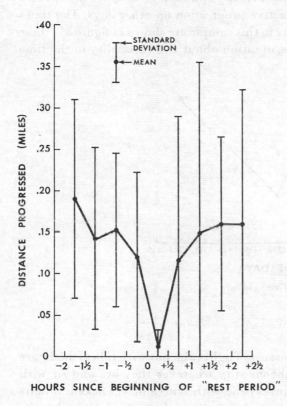

Fig. 34. Distance progressed by Main Group per half hour during afternoon progression minima and during preceding and succeeding time periods. See text for details.

group progressed the shortest distance, then aligned the remaining data for that day accordingly. A more exact description of the technique that was used is available from the authors.

As would be expected from the 'inertia' effect, progressions during the half-hour periods just before and just after the 'rest periods' tended to be slower than those at other times of the day. This effect can be seen in figure 34, in the data for distance progressed in the half-hour periods immediately before and after the minima.

We presume that the magnitudes of the minima are lower than would be expected from aligning the minima of days in which progression rates were determined independently for each interval, and according to the distributions shown in figure 24. Or to put it the other way around, we believe that figure 24 obscures the rest periods because they do not occur at the same time each day and their records are thus pooled with data from periods of active progression on other days. The times of the 31 progression minima in this sample are shown in figure 35; these data tend to confirm our assumption about the variability in the time of progression minima.

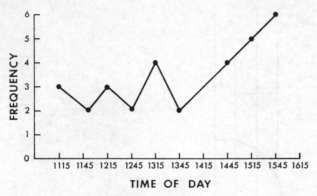

Fig. 35. Distribution of afternoon 'rest periods'.

7. Responses to Rain

Progressions do not cease because of light rain, but with rain of moderate or heavy intensity, the baboons stop wherever they are and sit with their backs hunched (i.e. convex) and with arms and legs flexed. Infants huddle in their mother's ventral flexure and are held there by the mother. In a strong wind, the animals generally sit with their backs

to the wind. We once observed a single adult male move 80 ft. away
from the group to the shelter of a tree, beneath which he huddled during
the ensuing rain while his companions remained in the open. CROOK and
ALDRICH-BLAKE [1968] write: 'In heavy rain *P. a. doguera* were seen to
climb into trees with a good canopy and sit still until the storm moderated.'

8. *Aggregation and Dispersion*

The extent to which the members of a group are spaced varies enormous-
ly. At one extreme, the Main Group was noted as being spread in an
irregular line for $^1/_5$ mi. on 2 occasions, $^1/_4$ mi. on 6 occasions, $^3/_{10}$ mi. on
3 occasions, and once for $^1/_2$ mi. (These measurements were facilitated
by means of a $^1/_{10}$-mile odometer, which we had installed on our vehicle.)
At the other extreme, the members of a group sometimes aggregated
into a small, roughly elliptical area. A sampling of the axes of such
minimal areas for the Main Group is as follows (all measurements in
feet): 13×60, 15×40, 15×45, 15×50, 20×20 (twice), 20×30, 20×40,
25×45, 30×30, 30×60 (twice), 40×60, and 50×50.

The baboons of a group aggregated under the following circumstances: (1) during encounters with potential predators; (2) in response
to strong predator alarm calls from another group of baboons; (3) during
false alarms; (4) when closely approached by another baboon group or
by Masai cattle; (5) in areas of heavy undergrowth; (6) before going
through a critical pass in the foliage; (7) when on an unfamiliar route;
(8) in response to a spatially restricted resource, such as a water source
or the shade of a tree; (9) at or slightly before the beginning of a group
progression; (10) in the evening, just before ascending a grove of sleeping
trees, and less often just after descending in the morning; and (11)
during the morning and evening 'social hours'. The first 7 situations
are actually or potentially threatening to the group. Numbers 9 and 10
are times of critical 'decision-making'.

Conversely, the members of the group became more widely spaced
under the following circumstances: (1) in open terrain; (2) just after
compacting under threat, as if from 'relief' or 'rebound'; (3) when on a
familiar route that has no critical passes.

The extent of group dispersion during feeding and resting seemed
to depend to a considerable extent upon additional factors. Certainly,
all of the maximal group dispersions occurred when the group was

foraging in open terrain, but on the other hand the members of a group would be moderately cohesive when feeding on understory plants in wooded areas. During some afternoon rest periods the group was about as compact as when in their sleeping trees, i. e. within an area of about 40×120 ft., but at other times they were more widely dispersed.

The spread of the group during progressions may be estimated as follows. Assuming a steady progression rate, the spread of the group, from front to back, will be the progression rate times the amount of time necessary for them to pass a fixed point. For practical reasons, censuses were taken only when the group was in the long file formation. For 52 of the censuses taken on the Main Group, we have a record of the time that the census began and the time that it ended. The difference is the amount of time required for the group to pass a fixed point—the 'counting point'. The progression rate for each census was taken as the mean rate during the half-hour interval that included the mid-point, in time, of the census. Seasonal mean progression rates were used whenever these were based on at least 10 observations; otherwise, annual rates (fig. 24) were used. From these data, the spread of the group was estimated for each census. The mean spread during these censuses was 263 ft. (S. D. 206 ft.).

9. Lunar Periodicities

Nocturnal predators that rely on vision are probably better able to locate their prey when the moon is full than at other times. From our tent, we sometimes heard baboon groups giving alarm calls in the middle of the night, in much the same manner as they do when they see a predator during the day, and it seemed to us that such nocturnal barking was more prevalant when the moon was full. At night, baboons in trees may be safe from all predators except the leopard. A direct test of this assumption would have required overnight observations, which were prohibited in the Amboseli Reserve. Nonetheless, it seemed likely to us that if baboons are, in fact, more susceptable to attack at full moon, they may be awake and alert for much of the night at such times, even if a leopard is not actually present[35].

[35] HADDOW [1951, p. 361] has noted that captive groups of vervet monkeys in large open-air runs at Entebbe, Uganda, may show considerable activity on nights of bright moonlight, and he cites SANDERSON [1939] for similar observations on New World monkeys. According to STOLTZ and SAAYMAN [1969], 'reliable informants report that baboons from W troop have been seen to descend the kopje on moonlit evenings during conditions of drought to feed from marula trees *(Sclerocarya caffra)* near the hotel rendavels. Similar observations have been reported to us by Nature Conservation Officers in Rhodesia.' LOVERIDGE [1921] reports baboons drinking at 2 a. m.

If so, it seemed likely that their daytime activities would be affected. We therefore looked at lunar variations in some of the baboons' activities.

Dates of full moon were obtained from a British almanac. (While Amboseli is some 37° east of Greenwich, this difference in longitude would result in only an occasional, one-day discrepancy in the date of full moon.) The lunar month was then divided into quarters, as follows. The day of the full moon, plus the three preceding and three succeeding days, were taken as the second quarter (full moon). The next 8 days were taken as the third quarter (waning moon), and the 7 days after that, the fourth quarter (dark moon). The remaining days (variously 7 or 8) were taken as the first quarter (waxing moon). Data from all days in the study that were the same number of days away from a full moon were tabulated together, and the results were then pooled by lunar quarters.

The results are shown in table XVIII. While the data available for an activity in any one quarter are somewhat sparse, there seems to be a consistent deviation of the data during the time of the full moon. On the average, the morning progression of the Main Group was longer at that time than during any other quarter. By noon, they were farther from the previous night's sleeping grove, and at 1700 they were farther from the next night's sleeping grove than during any other quarter. In short, the baboons tended to avoid the vicinity of sleeping trees more

Table XVIII. Lunar periodicity in selected activities of Main Group

	Lunar quarters							
	I Waxing moon		II Full moon		III Waning moon		IV Dark moon	
	$\bar{x}$	n	$\bar{x}$	n	$\bar{x}$	n	$\bar{x}$	n
Linear distance at 1200 from previous night's sleeping grove	0.063	14	0.071	20	0.070	16	0.068	19
Linear distance at 1700 to that night's sleeping grove	0.023	10	0.028	16	0.019	14	0.025	11
Distance traversed between 0830 and 1200	4.647	6	5.983	16	4.941	11	4.974	12

Distances are in miles. $\bar{x}$ = mean, n = sample size.

during days of the quarter of the full moon than at any other time in the lunar cycle. Whether this effect is directly related to greater susceptibility to predation at that time, or has some other explanation, cannot be determined without prolonged night-time observations. It is certainly contrary to our initial supposition that fatigue resulting from nocturnal vigilence would reduce the extent of progression during the day.

10. Home Range

The wanderings of a group of baboons may at first seem to be random and without bounds. This is because during any one day's progression the members of the group will move through only a small part of their total range, and several days may pass with no significant repetition of the area covered. But prolonged observation reveals that they move within a fairly circumscribed area and often along habitual routes. It thus becomes necessary to accumulate data from many days of observation in order to build up a composite picture of the area occupied by the group.

The most straightforward method of establishing the total home range would seem to be to plot all locations in which the group was seen. The results for the Main Group are shown in figure 36. But the attempt to draw a boundary around such a 'tangled ball of yarn', raises many interpretive problems.

A more informative method is to show the rate at which daily progressions of the baboons took them into areas not utilized during earlier days in the sample. This technique may not only lead to more accurate estimates of the total area occupied by the group, but may also give an indication of the amount of time that is required for the baboons to cover given portions of their home range. The rate of range occupancy of the Main Group was calculated in the following manner:

For each successive day of observations we measured the size of the area in the day-range of the group that was not included in any previous day-range in our sample. (Incomplete day-ranges were also utilized.) Day-ranges were delimited as follows. A 'taut string line' was drawn around the record of each day's progression line ('day-journey') on the map, enclosing the record of just that day's progression. (Mathematically, this is the convex hull of the day's progression line.) These simple, closed curves surround the day-ranges. Map representations of day-ranges were superimposed in chronological order (starting November 2, 1963) and for each day of observation, we measured the area not included in any previously observed day-ranges.

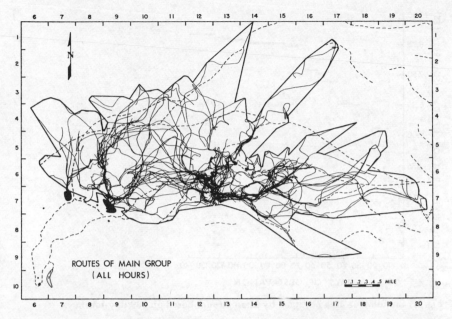

Fig. 36. Movements of Main Group. The map shows all observed pathways of the center of mass of the Main Group.

Cumulative totals of these area increments are shown in figure 37, based on 139 days. Even near the end of this long sampling period, the baboons sometimes entered new areas, though in general, this happened progressively less frequently as the study progressed. This suggests that the baboons have a relatively fixed home range beyond which they do not go, and that the size of this range could be estimated by the least upper bound of the distribution indicated by the points in figure 37. We have not attempted to estimate this least upper bound; the total area that we observed the Main Group to occupy was 9.299 sq. mi.[36].

11. Seasonal Changes in Range Size

We began our detailed study on the Main Group during October, 1963, which was the last month of the dry season. During that period, we began to work out the home range of the Main Group, using rough sketch

[36] This figure includes an area of 0.098 sq. mi. that was inadvertently omitted in making figure 37, but is included in all home range maps herein.

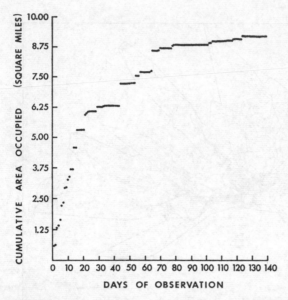

Fig. 37. Cumulative area occupied by Main Group, based on 139 day-journeys.

maps of the area—our detailed 1:15,000-scale maps were not completed until the following month—and during that early work we gained the impression of several sharp boundary limits to the group's range; i.e., there were several places on the periphery of the October range to which the Group repeatedly returned, but beyond which they did not go. On this basis, we felt confident that within a short time, we would have completely delimited the Group's entire home range. The area occupied by the Main Group during October is shown in figure 38.

Thus, we were astonished, in November, to find the group moving far beyond the limits of their October range. The onset of the rains on November 1 of that year brought about a major ecological change: thereafter, the baboons could get drinking water almost anywhere and thus were no longer tied to the vicinity of permanent waterholes. That this change in range size cannot be attributed to alterations in vegetation is evident from the fact that the change was apparent as soon as the rainy season began, during the first week of November, before the new grass had a chance to grow.

By the end of our study, in early August, 1964, no comparable reduction in the Main Group's home range had taken place. The group

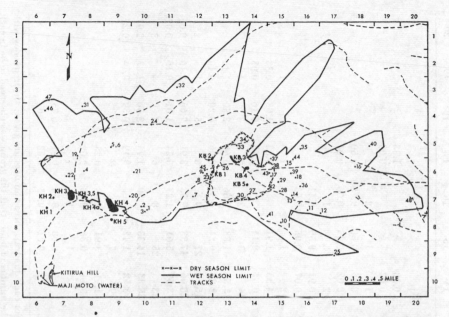

Fig. 38. Locations of permanent waterholes (numbers preceded by KH or KB) and major temporary rainpools (numbered dots). Home range boundaries are shown for the end of the dry season (October 1963) and for the rest of the study.

continued to move within an area that was many times the size of their October range (fig. 38). Possible reasons for this will be discussed in the next chapter.

12. Range Occupancy Distribution

The Main Group did not occupy the various portions of its home range equally. This can be shown in several ways, each of which attempts to depict occupation density. ('Density' is here used in the sense of probability, not individuals per unit area.) One way is to plot every place that the group was observed, as in figure 36. Another is to count the frequency with which each quadrat[37] of the home range was entered.

[37] Corners of the 0.4×0.4 mile quadrats that were used in this study were taken from fiducial crosses on an aerial photograph that was obtained from the Survey of Kenya (V13B/RAF/341, frame 158, 26 JAN 63).

Table XIX. Quadrat occupation data

Each cell gives three numbers: frequency (top), hours observed (middle), probability of occupancy (bottom).

Map row	Map column 6	7	8	9	10	11	12	13	14	15	16	17	18	19	20
1									1 0.58 0.07			1 1.33 0.09			
2						1 0.08 0.00	1 0.66 0.05	1 0.58 0.05	1 0.16 0.02	1 0.50 0.05	2 0.75 0.06	2 0.50 0.04			
3	2 0.25 0.03	2 0.66 0.04	1 0.58 0.06	2 0.83 0.06	5 6.83 0.84	7 5.66 0.64	4 2.33 0.19	2 0.75 0.07	2 0.91 0.07	3 0.50 0.05	1 0.41 0.03				
4	5 0.58 0.06	2 1.08 0.12	3 2.16 0.19	13 7.58 0.82	9 13.91 1.57	5 3.50 0.48	5 2.00 0.18	4 2.75 0.30	2 4.33 0.64	1 1.91 0.14		1 0.83 0.15	0 1.08 0.16	1 0.33 0.03	
5	1 0.58 0.09	2 3.41 0.51	17 8.91 0.90	18 16.16 1.75	7 4.16 0.45	10 9.91 0.97	11 7.00 0.76	20 21.50 2.45	14 15.58 2.23	3 1.83 0.21	2 2.66 0.37	2 0.66 0.06	3 2.50 0.20	4 1.16 0.10	
6	1 1.50 0.18	6 7.83 0.95	31 25.83 2.65	32 18.08 2.13	19 20.83 2.33	21 14.00 1.23	49 83.83 9.39	39 58.16 5.97	16 28.66 3.57	27 47.33 5.61	9 9.83 1.27	6 3.91 0.54	5 7.66 1.01	1 1.16 0.16	2 0.66 0.07
7		1 0.66 0.06	7 19.33 1.84	17 54.41 6.33	15 10.16 1.25	10 4.50 0.43	33 45.83 5.96	45 95.33 20.67	33 35.83 3.52	25 14.33 1.42	8 7.50 0.86	6 1.75 0.19	3 3.66 0.39	4 1.83 0.14	3 2.91 0.27
8				1 0.33 0.04	4 1.00 0.09	2 1.91 0.14			9 1.91 0.14	7 3.91 0.36	2 1.08 0.08	2 0.25 0.02			
9										1 0.08 0.01	1 0.50 0.03	1 0.91 0.08	1 0.75 0.07		

The rows and columns correspond with those on the maps (e.g. fig. 36). For each quadrat, three numbers are given: the frequency with which the Main Group was observed entering each quadrat (top number), the number of hours that the group was observed to be in each quadrat (middle number), and the probability of occupancy (bottom number) given as percent of time and corrected for the diurnal

Results of this second method are given in table XIX; the cumulative distribution is shown in figure 39. By either of these methods of depiction, it appears that a relatively small number of quadrats in the Main Group's home range account for a large portion of its quadrat entries. For example, from figure 39 we can see that just 11 of the quadrats accounted for 50 % of the observed entries.

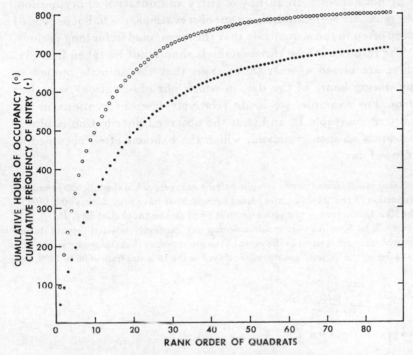

Fig. 39. Distribution of the frequency and duration of quadrat occupancy. There is one more quadrat on the upper curve because there is one quadrat in which the group was found one day, but which they were never observed to enter.

A third way to depict occupation density is in terms of the amount of time that the group was observed to spend in each quadrat. In mapping progression routes of the Main Group, the position at each half hour was marked on the progression lines. We have plotted the amount of time, between morning descent and evening ascent, that the baboons spent in each 0.4 × 0.4 mile quadrat, interpolating linearly to the nearest 5 min wherever a progression line fell in more than one quadrat. The results are 'mapped' in table XIX; the cumulative frequency distribution of observed occupation times is shown in figure 39. Note that this curve is even steeper than that for entry frequencies. For example, 50 % of the baboons' observed day time was spent in just 6 quadrats. The 42 quadrats in which the group was observed longest account for over 95 % of the group's observed time.

A particular quadrat does not necessarily have the same rank order by both criteria, frequency of entry and duration of occupation, but the ranks are correlated, i. e. in our observations, the baboons tended to return often to those quadrats that they remained in for long periods.

The results of these three methods should not be taken literally. All three are biased against those areas that the animals tended to occupy during hours of the day in which our observations were less frequent. For example, we made relatively few observations in the early afternoon (table I), and thus the observed distribution is biased against areas of open grassland, which the baboons often occupied at that time of day.

The bias that results from uneven sampling times was removed in the following manner. The proportion of the 'day' (i. e. the period for which we have map data, 0600–1900 h) that the Main Group spends in a given quadrat q will also be the probability, $P(O_q)$, that the group will be found in that quadrat during any randomly selected minute during the day, which in turn is equal to the mean of the probabilities that the group will occupy quadrat q during the various minutes of the day. For the 13 hours from 0600 to 1900, this mean is

$$P(O_q) = \frac{1}{780} \sum_{i=600}^{1859} p\,(O_q \cap t_i).$$

This can be estimated from

$$\hat{P}(O_q) = \frac{1}{780} \sum_{i=600}^{1859} \frac{n\,(O_q \cap t_i)}{n\,(t_i)},$$

where $n(O_q \cap t_i)$ is the number of times (days) that the group was observed to occupy quadrat q during minute i, and $n(t_i)$ is the total number of days on which observations were made at minute i[38].

To reduce the number of terms in the summation, we divide the day into 13 h, and assume that $n(t_i)$, the number of observations made at any given minute i, is the

[38] A minute is considered a sufficiently small period of time to be treated as an indivisible unit, in that during any one minute, the center of mass of the group can be considered to occupy just one quadrat.

same as that for any other minute during that hour of the day—and hence equal to $1/60$ of the number $N(T_i)$ of observation-minutes for that hour of the day[39]. Thus,

$$\hat{P}(O_q) = \frac{1}{780} \left[\sum_{i=600}^{659} \frac{n(O_q \cap t_i)}{N(T_6)/60} + \cdots + \sum_{i=1800}^{1859} \frac{n(O_q \cap t_i)}{N(T_{18})/60} \right]$$

$$= \frac{1}{13} \left[\frac{1}{N(T_6)} \sum_{i=600}^{659} n(O_q \cap t_i) + \cdots + \frac{1}{N(T_{18})} \sum_{i=1800}^{1859} n(O_q \cap t_i) \right].$$

In the last expression, the first summation, for example, is simply the number of minutes that the group was observed to spend in quadrat q during the hour beginning at 0600 h. Abbreviating that summation by $N(O_{q,6})$, and similarly for the other summations, we get

$$\hat{P}(O_q) = \frac{1}{13} \sum_{i=6}^{18} N(O_{q,i})/N(T_i).$$

The resulting probabilities, expressed as percentages of daytime, are given in table XIX. If the percentages of time in each quadrat are rank-ordered, and the contributions to total hours of occupancy made by each quadrat below any specified rank are graphed, the resulting curve is slightly above that shown in figure 39 for the uncorrected data. More important than this subtle difference in the graphs, however, is that the corrected estimates are not biased by our unevenly distributed observation times. As a result, the unbiased rank ordering of the quadrats is different. The 6 quadrats that were occupied the longest account for over half of the group's time; yet, these 6 quadrats have only 7% of the area of the 87 quadrats that the group is known to have entered, or about 10% of the home range area as estimated by the technique described on p.114.

About three-quarters of the Main Group's daytime and all of the nights were spent in only one-quarter (14 quadrats) of their home range area. These 14 quadrats form two areas, one in the KH woodland, one in KB (fig. 19). The obvious question is, what is special about these highly favored areas? These 14 quadrats, and some major natural resources provided by each, are indicated in table XX.

[39] Since we mapped the position of the Main Group during almost all times that we observed them, $N(T_i)$ is approximately the number of hours of observation on the Main Group during each hour of the day (B+C time, table I). However, because of some important exceptions (e.g. in October, 1963, before our field maps were completed), we have taken $N(T_i)$ to be $\sum_q n(O_{q,i})$.

Table XX. Resources of 'favorite' quadrats

Rank	Quadrat	Percent of time[1]	Sleeping groves	Percent of nights[2]	Waterholes Names	Percent of drinks[3]	Rain pools No. of pools[3]	Percent of drinks[7]	Major passes
1	(7, 13)	20.67	1, 10, 12	36.0	—	0	1	1.7	to 5 sleeping groves
2	(6, 12)	9.39	—	0	KB-1, -2	16.7	4	6.7	to 5 sleeping groves and 2 waterholes
3	(7, 9)	6.33	4, 5	13.6	KH-4	5.8	1	2.5	to 3 sleeping groves
4	(6, 13)	5.97	—	0	(KB-1, -2)[4], KB-3	10.8	1	0.8	to all KB waterholes
5	(7, 12)	5.96	2, 13	26.4	—	0	1	0	to 4 sleeping groves
6	(6, 15)	5.61	—	0	—	0	9	10.0	to KB woods
7	(6, 14)	3.57	3, 8	12.8	KB-4, (KB-5)	4.2	0	0	to 2 sleeping groves, 2 waterholes
8	(7, 14)	3.52	—	0	KB-5	8.3	2	4.2	to 3 sleeping groves, 1 waterhole
9	(6, 8)	2.65	11a, 11b	4.0	—	0	1	0	to 3 sleeping groves, 1 waterhole
10	(5, 13)	2.45	9	2.4	(KB-2, -3)	0	1	1.7	to 1 sleeping grove, 2 waterholes
11	(6, 10)	2.33	—	0	—	0	1	0.8	between KB and KH woods
12	(5, 14)	2.23	—	0	—	0	1	1.7	to 2 waterholes, 2 sleeping groves
13	(6, 9)	2.13	—	0	—	0	0	0	to 1 waterhole, 2 sleeping groves
14	(7, 8)	1.84	6, 7	4.8	KH-3.5, -4a[5]	1.7	0	0	to 1 sleeping grove
15–87	all others	25.35	—	0	KH-3[6]	1.7	25	22.5	

The essential resources provided by the 14 most-used quadrats of the Main Group are indicated. These 14 quadrats have about one-quarter of the home range area, but accounted for about three-quarters of the group's time.

[1] From data in table XIX.
[2] From data in table XI. On nights when the group split, they are arbitrarily counted as being in the grove with the lower number.
[3] From data in table XXI.
[4] Parentheses indicate that the waterhole falls into one or more other quadrats, and that most frequent use of the waterhole took place in one of these others. 'Percent of drinks' is based only on non-parenthetic waterholes.
[5] Grove 7 surrounds waterhole KH-3.5. The grove was used only once. Waterhole 4a is outside the known range of the group (see fig.38).
[6] Waterhole KH-3 was used just twice. It is in quadrat (7, 7), which accounts for only 0.06% of the group's time.
[7] Based on the 120 cases (out of 122 indicated in table XXI) for which the identity (location) of the rain pool is known.

The resources that are indicated in the table are sleeping groves, water sources and major travel routes or passes[40]. Food resources were not included. The baboons fed in virtually every quadrat of their home range. At present we have no measures of the quantity or quality of food that is available and utilized in the various quadrats. But none of these 14 favorite quadrats is primarily a feeding site. Each contains at least one of the other essential resources that are indicated.

From table XX we can see that the 14 favorite quadrats accounted for 75.8% of the Main Group's drinking sessions, and 100% of their sleeping sessions. They also account for virtually all major passes and corridors.

13. Diurnal Cycle of Localization

The distribution of the Main Group' sleeping groves was the major factor restricting the variability of th group's positions during the late evening, night, and early morning, at which times they were still in the trees. No other essential resource w ch a restricted spatial distribution is utilized during such a circumscri' ed (although long) period of the day. And similarly, their position, say, one hour after descent was limited to the area within an hour's progression from the trees. Positions during the hours just after descent, or just before descent, were further restricted by the availability of clear corridors or passes through the vegetation. Conversely, open savannah foraging areas are widely dispersed, and mid-day positions of the group were very variable. The resulting diurnal cycle of clustering of the Main Group's positions is illustrated by figures 40–43 for the hour intervals beginning at 8, 9, 11 and 17 h. The corresponding cumulative occupancy curves are shown in figure 44. (Occupancy data for other hours of the day are available from the authors.)

[40] By a *pass*, we mean a narrow preferred route of travel that results from the ecology of the habitat. A *corridor* is an elongated pass. For our baboons, these preferred passes and corridors seemed to be the shortest routes that connected sleeping groves, water holes and foraging areas, and that could be traversed readily, with a minimum of danger. In particular, open areas were generally preferred to areas with heavy undergrowth. The locations of these passes and corridors are clearly evident in figure 36.

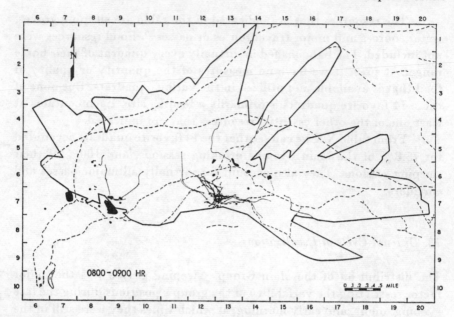

Fig. 40. Positions of Main Group between 0800 and 0900 h.

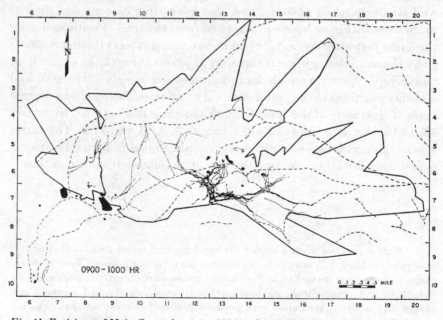

Fig. 41. Positions of Main Group between 0900 and 1000 h.

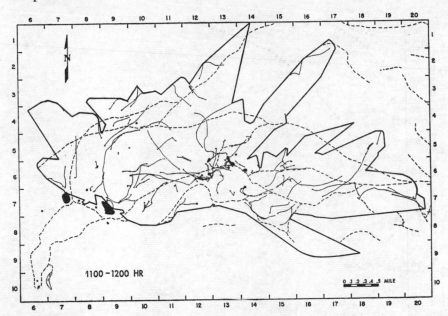

Fig. 42. Positions of Main Group between 1100 and 1200 h.

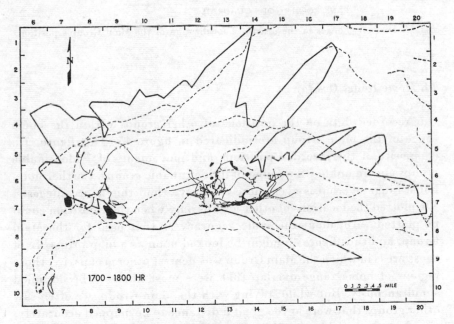

Fig. 43. Positions of Main Group between 1700 and 1800 h.

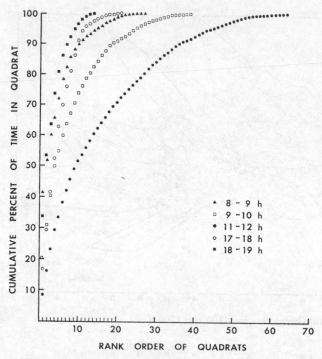

Fig. 44. Diurnal changes in the degree of localization of the Main Group's positions.

14. Home Range Overlap

Our recorded data on the positions of other groups within the home range of the Main Group are indicated in figure 45. This figure is a composite of the recorded positions and movements of 7 identifiable groups and an unknown number of unidentifiable groups. Furthermore. home range overlap is even more extensive than this figure suggests. We plotted the locations of other groups only when their position could be plotted with about the same accuracy as that used for the Main Group, and thus figure 45 might be looked upon as a map primarily of those areas in which the Main Group was nearest other groups, i. e. those regions of home range overlap that were most likely to be occupied simultaneously. But while staying with the Main Group, we often saw other groups that were at too great a distance to be mapped accurately, yet that were within the Main Group's home range. Our distinct im-

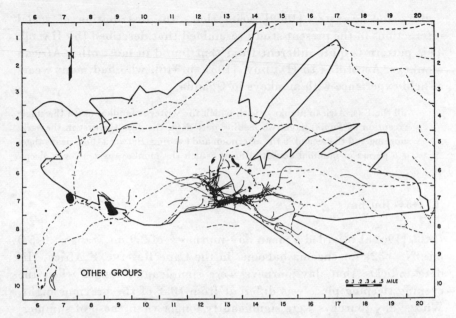

Fig. 45. Known location of other groups within the Main Group's home range.

pression was that there was no portion of the Main Group's home range that was not also a part of the home range of several other groups, or, to put it the other way around, there is no portion of the Main Group's home range that we believe to be occupied by them exclusively.

15. Comparisons and Discussion

(a) Diurnal Activity Cycle

The diurnal cycle of activity that we observed in yellow baboons was much like that reported by HALL [1963, p. 15] for chacma baboons:

> 'The diurnal pattern is characteristically made up of a period in the morning when social-sexual activity is greatest (at any rate in the Cape), with a second lesser peak near the sleeping place at the end of the day....
> Chacma groups, in all areas where we have observed them, do not rest at all regularly. Rather, their day has been one of persistent foraging for food, but with the occasional and irregular period of resting and grooming in bright sunlight.'

STOLTZ and SAAYMAN [1969] write that, 'the distribution of social
interactions in the present study resembled that described [by HALL]'.
This pattern is quite different from that found in most other African
monkeys. According to HADDOW [1951, p. 346], who had many years
of field experience with monkeys in Uganda,

> 'all the East African monkeys with which the writer is familiar (with the single
> exception of *Papio*) have two feeding-periods in the day, the first in the early
> morning and the second in the afternoon and evening. Between these peaks there
> is a period of minimal activity during which the monkeys rest or even sleep.'

(b) Day-Journeys

HALL [1962 a] recorded a mean day-journey[41] of 2.9 mi. (range 1.0–5.0
mi., N = 32) for chacma baboons in the Cape Reserve, S. Africa. His
data indicate that day-journeys were significantly longer when the
evening sleeping place was different from that of the previous night.
Winter day-journeys were significantly longer than those of summer,
'perhaps [because of] the greater concentration of suitable vegetable
food during the summer months' [HALL, ibid.].

In South-West Africa, HALL [ibid.] observed a group progress
12 mi. in one day. He comments: 'There is no reason to suppose that
this is a typical distance, but there is every likelihood from observation
of other groups in the area that day-ranges are necessarily longer in this
arid country of sparse vegetation than are those of Cape groups.'

In the northern Transvaal, STOLTZ and SAAYMAN [1969] recorded
a mean day-journey of 5.0 mi. (standard error 0.23 mi., range 1.5–9.0
mi.). They have shown that lengths of day-journeys in chacma baboons
of the northern Transvaal are inversely correlated with maximum daily
temperature; i.e., the baboons tended to travel farther on cooler days.
There was no correlation between length of day-journey and maximum
relative humidity. The distance did not vary markedly between summer
and winter, but neither did maximum daily temperatures.

KUMMER [1968, p. 168] gives a graph indicating mean progression
rates of hamadryas baboons at different times of day and at various

---

[41] HALL and some other authors would use the term 'day range' here, but we
prefer to reserve that term for the portion of the home range that the animals occupy
or utilize in one day, i.e., for an area, rather than a distance.

intervals after departure from the sleeping cliffs. His results are quite different from ours: hamadryas progress most rapidly between 8 and 9 in the morning, with a marked decrease in rate until mid-day, then an increase again in mid-afternoon. KUMMER's graph implies that hamadryas move a total of 13.9 km, on the average, between 0800 and 1600 h. This figure agrees closely with an average of 13.2 km per day (range 9.8 to 19.2) that KUMMER obtained on the basis of 8 daily routes that were known completely [ibid., p.167]; an exceptionally short route of 4.1 km was set aside. During this same period of time, 0800 to 1600, the Main Group of yellow baboons that we studied progressed 4.2 km (2.62 mi.), on the average.

Hamadryas day-journeys are, so far as we know, longer on the average than those reported for any other species of nonhuman primate. Other reports of unusually long progressions include HALL's one observation (above) of a 12-mile trek by a group of baboons in South-West Africa, BRAIN's report [1965] of a group of vervets moving 11 miles in one day in the Wankie National Park, Rhodesia, and FITZSIMONS' report [1919 quoted in ZUCKERMAN, 1932] of a population of chacma baboons in South Africa that migrated from one chain of mountains to another. In each case, these were exceptional journeys.

According to HALL [1963], 'the possibility of long-distance... migrations of baboon groups...could probably best be studied in S.W. Africa where prolonged drought is not uncommon, and where large-scale movement of baboons away from such regions as the Kaokoveld are known to occur'.

For other baboon populations, only brief comments have been made on progression rates and distances. For the forest-living anubis baboons that ROWELL [1966] studied in Uganda 'four miles was the longest [day's journey], a mile or a mile and a half typical, and on some days they moved only a few hundred yards.'

During one day, according to DEVORE and WASHBURN [1963], a troop of anubis baboons in Nairobi Park 'completes an average circuit of about 3 mi....but this distance varies from a few yards on some days to 6 or 7 mi. on others. These figures refer to the distance between points on a map. As a troop meanders across a plain, however, the individuals actually walk twice as far as these figures indicate.'

What is the adaptive significance of the baboons' use of preferred routes, corridors and passes? We speculate that baboons tend to route themselves so as to maximize their chances of survival in each situation, and that the preferred routes represent the most frequent solutions to these maximizing problems. The safety of preferred routes probably is enhanced by the great familiarity of the baboons with these areas. BERT *et al.* [1967a] have commented on the baboons' responses to even slight changes in such areas.

To illustrate, suppose that a group of baboons is moving in late afternoon from the open grassland to a sleeping grove. One available route leads through an open, fairly safe pass or corridor; the other, through a more restricted, riskier pass. However, the first route is longer, thus exposing the group to the dangers associated with being on the ground after dark, and increasing the chances that, in the end, all familiar sleeping groves will be occupied by other groups. The problem that the baboons face is that of minimizing their risks while maximizing their gains. The fact that the species is extant and, indeed, successful, indicates that in general, their solutions to these problems must be fairly good. But groups may differ in their ability to solve such problems; if so, differential mortality will result.

(c) Territoriality and Home Range

Attempts to clarify and sharpen our thinking about the land on which animals or groups of animals live have had an interesting development. Much of the early literature for mammals was reviewed by BURT [1943, 1949], who defined *home range* as that area normally traversed by an individual in its day-to-day activities of food-gathering, mating, and caring for young. More recent reviews of mammalian home ranges include those by JEWELL [1966] and BROWN [1966].

Territory, on the other hand, was defined as any defended part of the home range. This is essentially the concept of territory that was developed on the basis of bird data by ALTUM [1903] and independently by HOWARD [1907–1914, 1920, 1929.] The concept of territory as 'any defended area' [NOBLE, 1939] is now almost universally accepted.

There is at present no convincing evidence for territoriality, in the sense of a defended area, in any population of baboons, including those that we observed. Indeed, the presence of territorial defense in baboons has been specifically denied [HALL, 1962a; ROWELL, 1966; WASHBURN and DEVORE, 1961]. STOLTZ and SAAYMAN [1969], after describing intergroup antagonism in chacma baboons, write: 'For the present the significance of these observations for concepts of territorial defense in baboons remains obscure.' Apparently the only two suggestions in the literature of territoriality in baboons are by HALL [1965 b, *contra* HALL, 1962 a] and by MAXIM and BUETTNER-JANUSCH [1963]. The latter write: 'We cannot argue strongly that our observations are direct evidence for it [territoriality].' The cases that they and HALL describe may just as readily be attributed to defense of group integrity.

Aside from the question of territoriality, two major problems arise repeatedly in studies of home range: how to delimit the boundaries of the range and how to describe differential utilization of areas within this boundary. In studies of mammalian home ranges, these problems are confounded by the fact that much of the work is based on capture data. For many species of small mammals, direct observations are not feasible. Several techniques have been proposed for systematically demarcating the home range from such data [MOHR, 1947; STICKEL, 1954]. DALKE [1942, also DALKE and SIME, 1938] drew straight lines connecting the extreme outermost points at which the animals were observed or captured. The resulting polygon was referred to as the *minimum home range.* On the other hand, ODUM and KUENZLER [1955] refer to this polygon as the *maximum home range,* and believe it will often be larger than the *utilized home range,* that is, the portion of the maximum home range that the animals actually use.

Some authors, e.g. DICE and CLARK [1953], believe that for mammals without territories there is no fixed limit to their wanderings and thus attempts to fix boundaries on home ranges and to express home ranges as areas are futile. Others, including DASMANN and TABER [1956] and KAUFMANN [1962] felt that the mammals that they had studied had home ranges with definite limits even though there was no defended territory within that range.

In studies of differential use of land within the home range, several authors have attempted to distinguish areas of special significance (other than territories). PITELKA [1959] pointed out the ecological significance of exclusively occupied portions of the home range. He proposed calling such areas *territories,* but to do so would only result in confusion over this well-established term. These exclusively occupied areas are referred to as *monopolized zones* by JEWELL [1966].

KAUFMANN [1962] defined the *core area* as 'the area of heaviest regular use throughout the entire period of observation... determined by inspecting the daily route maps for each band.... Areas of heavy temporary use...were not included.' Thus, the core area is essentially the same as the *focus* of the home range, which was defined by CARPENTER [1934] as 'the place where the group was found most frequently'.

The concept of core area has been utilized in several primate field studies [e.g. DE VORE, 1962; JAY, 1965]. An examination of actual activity records and the core areas that have been drawn from them [e.g. KAUFMANN, 1962, p.172; DE VORE and HALL, 1965, p.35] clearly reveals the amount of arbitrariness that is involved in drawing a boundary around the proposed core area.

Of course, distinguishing the *most* frequently used area from the *least* is just a first approximation. A few attempts have been made to use the loci of observations or of trappings to estimate the animal's actual distribution of land use. As a result of DICE and CLARK's speculation [above] that nonterritorial mammals have no fixed range boundaries, these authors proposed that home range be expressed in terms of an *activity radius*, based on the distance of recaptures from the *recapture center*. The recapture center concept was apparently first used by HAYNE [1949] and called the *center of activity* by him. It is the geographic center of all the points of observation or recapture; that is, it is the point whose coordinates are the means, on the x- and y-axes, of the points of capture. As such, it is the point with the smallest total squared distance to the points of capture as well as the center of gravity of those points. Around this center, HAYNE visualized concentric rings or zones of equal probability of capture.

On the assumption that the animals' use of their home range has a bivariate normal distribution which centers on the recapture center, one can obtain expressions for the probability of finding or trapping an animal in any part of its range, the probability that two animals will simultaneously use overlapping portions of their range, and so forth [DICE and CLARK, 1953; CALHOUN and CASBY, 1958; JORGENSEN, 1968]. CALHOUN [1964] has pointed out a number of population consequences of home ranges with such distributions.

In our study, we had the advantage of direct observations on the animals' positions and activities. While it would be possible to use these observations to determine a position center, analogous to the above-mentioned recapture center, the observed home range of our animals was not at all circular, nor was the distribution of activities within that area radially symmetric. It appears that no such simple geometric interpretation of differential activity within portions of the home range will suffice for baboons.

Our major reason for looking closely at the distribution of movements within the home range is to understand how the animals utilize this land in order to survive. The home range is not a homogeneous piece of land. Different parts of the range have different things to offer. The most straightforward approach to the problem of preferential land use is to observe how the animals distribute their activities among the various portions of their range, and record what each portion provides the animals. The biological relevance of differential land use can then be understood by relating it to the distribution of resources and hazards within the area.

We sympathize with JEWELL [1966] who wrote: 'It would seem that further attempts to define the qualities of home range in a formal way are not very useful at the present time, when so little is known about home range behavior. What is required is an understanding of the ways in which an animal makes use of the terrain in which it has established itself.'

There have been a few previous attempts in primate field studies to establish the distribution of time among quadrats of the home range. These include the work of STRUHSAKER [1967] on vervets, STOLTZ and SAAYMAN [1969] on chacma baboons, and ROWELL [1966] on anubis baboons, none of whom have attempted to remove the bias that is introduced by the distribution of observation time.

We have indicated the concentration of water sources and sleeping trees in those portions of the home range that were most frequently used. Similar results were obtained on chacma baboons by STOLTZ and SAAYMAN [1969]: '...the areas most frequently occupied by the troop were situated in the vicinity of water reservoires...the vicinity of the sleeping cliffs represented a secondary focus of troop activity. Hours were evenly distributed throughout the major vegetation types.'

STOLTZ and SAAYMAN [1969] have indicated that seasonal shifts in the portions of the home range that were utilized by the chacma baboons that they studied in the Transvaal may have resulted from the seasonal availability of cocoons of the sweet-tasting mopane fly and from the distribution of available drinking water. They emphasize that the latter also had marked effects on the choice of sleeping sites, on home range size and, as noted above, on the length of day-journey.

In ROWELL's study, which was conducted on baboons in the Ishasha River valley,

'the time spent in forest was very high: in the home ranges of S and V troops, 18 % of the grid squares contained forest, river edge bush, or forest edge, but the animals spent 60 % of the time in them.... Vegetation types were associated with different activities: open sandy areas, and open places on the river bank, were favoured as resting places; foraging was seen mainly in short or course [coarse ?] grass. Isolated trees were typically eating places, abandoned as soon as the troop was satiated. In forest, and high bush, the pattern was eating and resting, followed by a very short walk to the next site. This pattern, also reflected in the short daily journey, is different from that of open country baboon troops, which seem to keep moving fairly steadily through the day, foraging as they go.... Each troop concentrated much of its activity in a very small part of the available area. These could perhaps be regarded as "core areas" of the ranges, but they were not the exclusive "property" of any one troop. There was a suggestion, though no good evidence, that S and V troops avoided each other in their area of overlap in so far as each troop usually used it when the other was at the far end of its range.'

(d) Home Range Size

DEVORE and HALL [1965] give home range sizes for 8 groups in Nairobi Park, Kenya, and 3 groups in Cape Reserve, South Africa, and comment that 'the ranges for only 4 of these groups is judged to be complete'. According to their table 2–4, these four groups are groups LT and SR of Nairobi Park, and groups N and S of Cape Reserve. For groups SR and S, samples of 55 and 10 day ranges, respectively, are indicated. If we assume that the relative rate of area increment with continued sampling is about the same in these baboons as in the Amboseli baboons that we observed, then the home ranges estimated for these two groups are, respectively, 82 % and 35 % of what they would have been had these groups been sampled for 139 days, as in our sample, and thus are presumably even smaller percentages of the actual total range. For the remainder of DEVORE and HALL's groups, the number of day-ranges that were sampled presumably was, in general, smaller (though a sample of 21 day ranges is indicated for one group). On the same basis, DEVORE and HALL's estimate, that the home range 'sizes shown for the other 5 [7 ?] groups probably include 80 % of their annual range', is too high.

We have indicated that no apparent reduction in the Main Group's home range had taken place by the end of our study, in early August, 1964, despite the fact that the dry season was well underway and the baboons got free water with decreasing frequency from sources other than permanent waterholes. Several possible explanations for this have occurred to us. Perhaps the small October, 1963, range was a sampling error, or a response to the constant presence of then-unfamiliar observers. We doubt both of these possibilities and suggest that an ecological explanation, related to water balance, is more likely. One might suppose that the home range is restricted to the vicinity of permanent waterholes whenever the need for free water is very great and the permanent waterholes are the only available source. If so, the baboons would be expected to drink more frequently during the dry season than during the rainy season. But exactly the opposite seems to be the case (p. 143), so this suggestion must be discarded. Alternatively, baboons may considerably reduce their water loss by remaining in the forested areas that occur near permanent waterholes. For instance, going from the KH woods to the KB woods meant crossing large, un-shaded areas of open grassland. Yet, much unused woodland was virtually continuous with the portion of the KB woods within which they remained during October.

We shall continue the discussion of the effects of food and water resources on home range in the next chapter. In the final chapter, we shall return to the questions of how baboons sustain themselves within a relatively fixed area, how they allocate their time and energy among the various portions of their home range, and how the land is distributed among the various groups.

Corrigendum added in proof. On the basis of discussions during 1969 with aerial photographers at the Survey of Kenya, the scale of the 1963 aerial photographs is probably closer to 1:12,500 than to 1:15,000, as indicated on p. 11; however, ground control is still not available for Amboseli. If the scale of 1:12,500 is correct, then all linear measurements for Amboseli that are given in this monograph should be reduced to $5/6$ of the stated values, and all area measurements should be reduced to $25/36$ of the stated values. All distribution parameters (mean, standard deviation, etc.) may be similarly scaled down (because expected values are invariant under linear transformation of the random variable). During 1967, new, clearer aerial photographs of Amboseli were taken (V13B/RAF/585, frame 014, 17 JAN 67, and V13B/RAF/616, frame 116, 2 FEB 67).

Corrigendum added 1973. During 1972, S. Altmann obtained ground control by measuring, on the ground and on the aerial photograph, the sides of a triangle that was on virtually flat, level ground and that was demarcated by three bushes that appeared as small dots on the photographs. These measurements give estimated scales of 1:14,842, 14,588, and 14,853 (Mean 1:14,761).

VI. WATER AND FOOD

1. Sources of Water

Two major sources of free water are available to baboons in Amboseli, permanent waterholes, fed by springs[42], and temporary rain pools. During our study, the Main Group is known to have utilized 40 of the 48 rain pools within their home range[43]. All waterholes within the Main Group's range were used[44], some (especially KB-1, KB-3, and KB-5), more frequently than others (see table XXI). Differences in frequency of use seemed to be due primarily to proximity of 'favorite' waterholes to the group's major progression routes, rather than to the absence of surrounding vegetation that might conceal a predator. For example, waterhole KB-1, around which vegetation is fairly heavy, was used more frequently than KB-2, the south and east sides of which are relatively free of obscuring vegetation. However, when baboons and vervets did use KB-1, they were observed drinking only from the sides where the foliage is least obscuring.

When a group of baboons goes to a waterhole, they do not all drink at the same time, even at those water sources for which shore access is not a limiting factor. Rather, a few drink and then move away, to be replaced by others. At times, members of the group remain in the vicinity (fig. 46), and a few may climb nearby trees. At other times, the baboons file past the edge of the water, each baboon drinking when at

[42] The areas around the largest of these springs are referred to as *swamps* on government maps; there is similar vegetation, on a smaller scale, around the large waterholes.

[43] The exceptions were: rain pools 4, 7, 14, 39, 41, 46–48. Of course, immediately after a heavy rain, there are puddles in many places. We have tried to distinguish the largest pools, as well as those used by the Main Group.

[44] Waterhole KII 3.5 is a marginal case. Baboons of the Main Group were observed in the grove of trees around this small waterhole on the morning of June 27, 1964. The observations were made by Dr. IRVEN DeVORE and his associates. We did not locate the group until 0915, at which time it was already on the ground, and about 80 ft. from KH 3.5. We do not know whether this waterhole was used by them. At no other time were they observed at this water source.

Table XXI. Diurnal changes in use of individual waterholes and of rain pools. The table gives numbers of drinking sessions in the Main Group

	Waterholes										Rain pool	Total
	KH 3	KH 3.5	KH 4	KH 4a	KH 5	KB 1	KB 2	KB 3	KB 4	KB 5		
6	0	0	0	0	0	0	0	0	0	0	0	0
7	0	0	0	0	0	0	0	0	0	0	0	0
8	0	0	0	0	0	1	0	0	0	0	1	2
9	0	0	1	0	0	2	0	1	1	1	8	14
10	1	0	0	0	0	3	1	4	1	1	12	23
11	0	0	2	0	0	2	2	3	0	4	13	26
12	0	0	0	0	0	1	0	1	2	1	8	13
13	0	0	1	0	0	0	0	1	0	0	4	6
14	0	0	0	0	0	2	0	2	0	0	4	8
15	0	0	1	0	0	1	0	0	0	0	3	5
16	1	0	1	0	0	3	0	0	0	1	3	9
17	0	0	0	0	0	2	0	1	1	2	7	13
18	0	0	1	0	0	0	0	0	0	0	2	3
19	0	0	0	0	0	0	0	0	0	0	0	0
Total	2	0	7	0	0	17	3	13	5	10	65	122

Time of day (beginning of interval)

the shore. These procedures probably reduce the risk of being surprised by a predator, since there is never a time when all the animals are drinking, and at all times some of them are near the drinkers, looking about.

When shore access is limited, as at a partially dried-up pool, dominant animals sometimes supplant subordinates from the drinking spot, but we never observed intense agonistic conflicts in those situations. The dominance relations are apparently established at other times and places. KUMMER [1968, p. 122] writes of hamadryas baboons as follows:

'With some exaggeration one might say that in the safety of the sleeping rock, social relationships are being established, whereas in the open field they are in function; here, their ecological value comes to the test, and during this subtle functioning the purely establishing patterns of behavior are suppressed.'

Fig. 46. Baboons at waterholes and temporary rain pools.

a

b

c

2. Frequency of Drinking

How often do baboons drink? How long do they go without water? Do baboons drink every day? Altogether, we observed 122 drinking sessions [45] in the Main Group during a total of 1,006.12 daylight hours, or an average of 0.12 drinking sessions per hour of observation. Intervals between drinks can be obtained from 'complete days', i.e. days with complete drinking records. Such records are available for 19 days [46].

Of these 19 days, there were 3 on which the baboons had no drinking session. Thus, the baboons went without water on 16% of the days in the sample. There were 11 days (58%) on which they took one drink, and 5 (26%), on which they took two. When the group drank but once, mean time of the drinking session was 1313 h. When the baboons drank twice, the first drink was at 1026 h, the second, at 1346 h, on the average; thus, on such days, there was an average of 3 h and 19 min between drinks.

The 19 days for which we have complete drinking records include four runs of two or more consecutive days. The numbers of drinks per day in these runs were as follows:

	Day			
	1st	2nd	3rd	4th
Run I	2 →	1 →	(1) →	1
Run II	1 →	2 →	0	
Run III	1 →	(1) →	1 →	1
Run IV	2 →	2 →	(1) →	1

Parentheses indicate drinks observed on single days with incomplete drinking records where such days join a run of 'complete days' to another 'complete day'.

The observed drinking sessions of the Main Group were distributed through the hours of the day as shown in table XXII. However,

---

[45] See footnote 47, p. 142.

[46] This includes 12 days on which we observed the Main Group continuously, from before they descended the sleeping trees in the morning until they ascended again that evening, and 7 days on which we have complete observations except for a short period after descent or before ascent, during which we are confident that the group did not have access to a rain pool or waterhole.

Table XXII. Seasonal changes in drinking sessions of Main Group

		Jan-Feb WH RP		Mar-Apr WH RP		May-Oct WH RP		Nov-Dec WH RP		Total
	0700	0	0	0	0	0	0	0	0	0
	0800	0	0	0	1	1	0	0	0	2
	0900	0	1	1	2	5	1	0	4	14
	1000	0	0	0	5	9	5	2	2	23
Time of day (onset of interval)	1100	3	4	0	3	10	3	0	3	26
	1200	0	0	0	3	4	2	1	3	13
	1300	0	0	0	2	2	2	0	0	6
	1400	0	0	1	2	2	2	1	0	8
	1500	0	1	0	0	2	1	0	1	5
	1600	0	0	0	3	6	0	0	0	9
	1700	0	1	0	5	5	0	1	1	13
	1800	0	1	0	0	1	0	0	1	3
	Total	11		28		63		20		122
	Drinks per hour of observation	0.13		0.12		0.11		0.16		0.12

WH waterhole. *RP* rain pool.

this sample is biased because our observations were not evenly distributed through the day. For each hour of the day, the probability of drinking at least once was estimated from the number of days with at least one drinking session divided by the number of hours of observation at that hour. The results are shown in figure 47. There is a peak between 1100 and 1200 in the probability of drinking, with a fairly high probability maintained for 6 h thereafter, i.e., during the hottest part of the day or as soon thereafter as the group got back to water after foraging.

That the number of drinks per hour of observation during any (hour) interval is an estimate of the probability of drinking during that interval can be seen from the following. Data for the interval come from two kinds of days, those on which we observed throughout the interval and those on which we observed during only part of the interval. An estimate could be based solely on the former by dividing the number of days on which a drink was observed by the total number of days in the sample, i.e., the number of successes divided by the number of successes plus failures. Days with data for only part of the hour can also be included in this calculation if we assume that the probability of drinking is uniform throughout the hour, and thus that the contribution of each such day to the numerator and denominator of the estimate is proportional to the fraction of

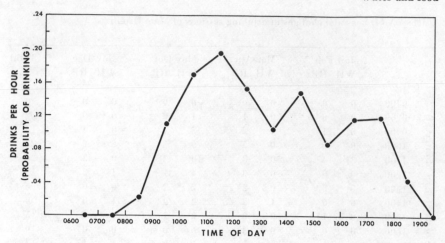

Fig. 47. Probability of drinking by the Main Group during each hour of the day.

an hour during which observations were made. If this is done, the resulting estimate is obtained by simply dividing the total number of days on which the group was observed to drink during the hour interval by the total number of hours or fractions thereof during which the group was observed during the interval.

3. Seasonal Differences in Drinking

Although the timing and quantity of rainfall in Kenya varies from year to year, there are typically two rainy seasons, the first in November and December, the second in March through April or early May. Between these two rainy seasons are two dry seasons [GRIFFITHS, 1962]. During the year of our study, the January-February period was never totally dry: rain came often enough to maintain green grass and to keep water in the larger, seasonal rain pools.

The number of drinking sessions[47] by the Main Group that were observed during these four seasons are shown in table XXII. However,

---

[47] The term 'drinking session' refers to a group activity, in which numerous members of the group drink from a water source. It does not include times when only one individual drinks. Such solitary drinking is particularly frequent during the rainy seasons, when many small puddles of water are available. Thus, the higher frequency of drinking during the rainy season is even more marked than the above data indicate.

these frequencies are biased by differences in observation time in the four seasons. In each season, an estimate of the probability of drinking can be obtained, as before, from the number of drinks per hour of observation on the group (B + C time, table I). The results are shown in the last row of table XXII. Contrary to what one might expect, the animals were most likely to drink during the heavy rains of November-December and least likely during the long dry period, from May to October. Drinking was about as likely during the January-February inter-rain period as it was during the short rains of March-April.

In the preceding chapter, we described the fact that in October, 1963, at the end of the dry season, the Main Group had a range that was limited to the vicinity of the KB waterholes, that this range increased dramatically with the onset of the rains in November of that year, but that no reduction to the October range had taken place by early August of the next year. This use of areas remote from permanent water may be directly related to the weather and to the continued availability of water in the largest temporary rain pools throughout this 'dry' season.

There was heavy rain in Amboseli during the last week of April, 1964, at which time the pools were filled. There was no rain during May, but on many days, there was a heavy overcast. During that month, the Main Group used at least 10 rain pools (No. 3, 6, 8, 15, 20, 21, 22, 31, 32, 40). By the first week of June, there was still no rain, the grass was almost entirely brown, and the rain pools were dry or almost so. On June 5, we estimated that the muddy water in one rain pool that was used by the Main Group (rain pool No. 5) would not last another week. But heavy rain fell on June 8, thus replenishing the pools. At least four rain pools (No. 5, 20, 24, and 40) were used that month. Weather during the next two months was generally cool and the sky was often heavily overcast. Such weather doubtless cuts down on evaporative loss from rain pools. Nevertheless, by July 2, rain pool No. 44 was, so far as we know, the only temporary rain pool left in the Main Group's range that still contained potable water. Nine days later, there was only mud in the bottom of it. On July 12, the Main Group moved to rain pool No. 3, only to find it dry; so, too, were all the other rain pools in that area. We have no record of any rain pool being used during July or the first week of August and the baboons were completely dependent upon the permanent waterholes. Nonetheless, the Main Group's range remained large. Perhaps the cool weather and heavy overcasts during that period reduced their water requirements.

4. Plant Food

Baboons have a very diverse diet. Their lack of highly specialized anatomical adaptations, combined with their keen vision and their ability to climb, dig, pull, pluck, gnaw, and to move great distances, often far from trees, enable them to exploit a wide variety of foods, and thus to survive in a great diversity of habitats. Baboons are the most widespread and abundant nonhuman primates on the African continent.

Of particular importance in the diet of Amboseli baboons are grasses and acacia trees. Virtually all of the trees in the study area are of one species of acacia, the fever tree (*Acacia xanthophloea*); in the drier parts of the study area, fever trees are replaced by umbrella trees, *Acacia tortilis* (fig. 2). Many parts of these two acacias are eaten: leaves (rarely), sap, cambium (when exposed by elephants), blossoms, seeds, seed pods, and even rotten wood (perhaps because of termites or other insects).

The gummy sap of the fever tree was, to us, both odorless and flavorless, yet the baboons seemed attracted to it, and would even leave a group progression if one of them sighted a sap ooze.

Fever tree seeds were eaten in great numbers. Much ground foraging beneath fever trees was initially mistaken by us for grass eating, but closer observations revealed that in many such cases the baboons picked out the acacia seeds that had fallen into the grass. At close range we could readily hear the crunching of the seeds and smell their garlic-like odor.

Acacia blossoms and green pods were available only during a small part of the year, at which time they were eaten in large quantities. Many fever trees in Amboseli were in full bloom during mid-October, 1963. We recorded acacia-flower eating between 12 October and 2 November 1963, with a peak about 18 October. On that day individual feeding rates in adults averaged 10 'bites' (one or more blossoms per bite) every 16.7 sec (sample of 17, 10-bite sessions). Such feeding may continue for several hours a day; thus, enormous numbers of blossoms are consumed. Green acacia pods were eaten, beginning the first week of November (fig. 48). A month later, the pods were drying and losing their green color. Dry pods were eaten, too, though perhaps not as often as green ones. The dry pods were taken from the trees or from where they had fallen on the ground. By May 1, however, very few pods still hung in the trees.

Many of these parts of acacias, particularly the pods and seeds, are good sources of protein. DOUGALL, DRYSDALE and GLOVER [1964], on the basis of their analysis of Kenya's browse and pasture herbage, indicate

Fig. 48 a) and b). Feeding on green pods of fever trees. *c)* Feeding on sap of fever tree.

a

b

c

that the legumes are the richest source of protein and are the least fibrous of all the vegetation.

Grass seeds were eaten when available. Particularly favored were those on stellate seed heads, such as *Dactylotenium* sp. nov. and *Cynodon plectostachyus*. The baboons neatly stripped the seed heads between their fingers, or between their tongue and teeth. Such selective feeding on grass seeds during dry periods means that the baboons are using one of the most nutritious parts of the plant.

Grass blades were commonly eaten when young and green (i.e. after the onset of the rains), but only rarely when dry. On November 22, just three weeks after the onset of the torrential 'long' rains, a major change in the baboons' feeding habits took place: an estimated 90% of the bulk of their food consisted of green grass that had grown since the onset of the rains. Such new growth was eaten by infants, as well as by adults (fig. 49). DOUGALL, DRYSDALE and GLOVER [1964] point out that

'the leafy tissues of young grass are rich in protein and minerals and they are not fibrous.... As grass ages physiologically, its percentage content of protein and minerals declines as these components become progressively diluted in an increasing percentage content of non-nitrogenous organic matter, including crude fibre; its nutritive content declines concomitantly....'

The baboons' 'specialty', particularly during the dry season is the rhizome and base of the blades of grass plants. They liberate the rhizome by digging around the grass plant with their hands, then pulling the plant up, using the hands or the mouth (fig. 50). In this way, an adult male baboon may consume 8 grass plants in 10 min, as timed observations revealed. When rain has softened the earth, grass plants are sometimes pulled up with little or no digging, particularly by adult males. The plant is often pulled out with the teeth. After a grass plant is pulled up, the baboons often wipe or rub dirt from the roots and rhizome, a technique that probably reduces wear on the molar teeth [DEVORE and WASHBURN, 1960]. They then eat the fleshy rhizome and particularly the white, subterranean bases of the grass blades, but little or none of the dry, upper stalks.

In utilizing the rhizome, baboons are getting at one of the richest sources of nutrients in the grass plant; they are tapping a major food and water supply that is not available to grazing ungulates.

Two grasses deserve particular comment. *Sporobolus robustus* is a tall, tough grass that grows on the edges of pans. There it forms dense

Fig. 49. Infant eating fresh, green grass at beginning of rainy season.
Fig. 50. Adult female feeding on rhizome of grass plant. Note full check pouches.

stands that would easily conceal a predator. Baboons were often alert
or tense when near this grass, particularly when they passed through it.
They seldom ate it (10 observed cases in the Main Group), and so far
as we could tell, it was seldom eaten by other animals. Another grass,
Cynodon dactylon (or *Sporobolus marginatus*?) grows as a thick mat
around waterholes. Despite numerous trips to the waterholes, baboons
of the Main Group were not observed feeding on this plant, except
during the period October 23–November 4, when they fed extensively
on it. The new growth of *Cynodon dactylon* is high in protein [Dougall
and Glover, 1964].

Plants other than acacias and grasses probably contributed less
to the baboons' total diet, though some of these were seasonally very
significant.

The small, red fruit of *Solanum incanum*, was eaten between
February 11 and July 30; several times during that period (e.g. March 2,
March 13, June 23, July 10) members of the Main Group gorged themselves on these berries, which to us had a most unpalatable, acrid-bitter
flavor. The leaves of this species are high in protein, calcium and phos-
phorous [Dougall, Drysdale and Glover, 1964], but the fruits apparently have not yet been analyzed. The genus *Solanum* includes the
deadly nightshade and several other toxic plants. No adverse effects
were noted after the berries were eaten by the baboons. A species of
Solanum that grows in Uganda is avoided by the baboons there [Ro-
well, 1966].

At times, a succulent, prostrate plant, *Trianthema ceratosepala*,
was a common source of food (feeding recorded June–October). Sometimes the edge of the plant was rolled over, like a rug, thereby exposing
the buds, which were picked off and eaten. The leaves have a slightly
salty taste, but are otherwise flavorless.

The gray-white berries of a common spiny bush, *Azima tetracantha*, were eaten; feeding was most extensive during March 8–12. New
leaves were eaten occasionally from the last week of May until mid-July.

Papyrus plants *(Cyperus laevigalus* or *C.immensus)* from the
waterholes were occasionally eaten by the baboons.

Leaves of *Salvadora persica* were once observed (by Thomas
Struhsaker on May 7) being eaten by baboons (not from the Main
Group). We have no other record of feeding on this common plant. It
has no outstanding nutritive value except as a source of calcium
[Dougall, Drysdale and Glover, 1964]. According to Napier Bax

and SHELDRICK [1963], this plant was eaten by elephants in Tsavo
Park, Kenya, during a period of drought (1961) but not at other times.

   On April 21, new growth on *Suaeda monoica* was eaten, but otherwise this common plant seems to have been ignored by the baboons.
DOUGALL, DRYSDALE and GLOVER's sample [ibid.] of twigs, blackish
mature leaves, and flowers of this plant indicates a remarkably high
sodium content—by far the highest of any plant that they analyzed—
but no other outstanding nutritional characteristics.

   A monocot resembling lilies springs up in some grassy areas after
the onset of rains. We observed baboons eating it twice, March 15 in
the SW portion of quadrat (5, 9) and May 7 in the NE portion of quadrat
(7, 11); on the latter date, it was beginning to flower.

   On one occasion (December 21), a juvenile-2 was observed eating
white mushrooms.

   Several unidentified and unfamiliar plants were eaten occasionally. The cherry-sized green fruits of a vine that climbs on fever trees
were eaten on March 11 by one subadult male, the leaves, on May 24
by one juvenile male. Green seed pods of a plant *(Commicarpus pedun-
culosus?)* with a small, yellow flower, growing beneath umbrella trees
in the SW corner of quadrat (5, 19), were eaten by many members of
the Main Group on May 25. On July 30, a mature female ate flower buds
of a vine-like plant growing in depressions near *Solanum incanum*; the
flowers are tiny and white, and grow in clusters. As she ate, she was
watched by other baboons of the Main Group. The seed pods or berries
of one other unidentified plant that grows in these depressions was also
eaten. That exhausts the list of known food plants of baboons in Amboseli.

   Leucas stricta, one of the few plants that grows on certain patches
of a distinctive type of soil in Amboseli, was not eaten by the baboons.
Upon entering such an area, the baboons stopped feeding, traversed
the patch, passing numerous plants of this species, and then resumed
feeding on the other side. The nutritive content is not known.

5. Animal Food

Baboons sometimes prey on other animals. Grasshoppers were, by far,
the animals preyed upon most frequently by baboons of the Main
Group. A quick swat at a grasshopper on the ground was sometimes all
that was necessary to catch it. At other times, grasshoppers that took

flight were chased and sometimes caught in mid-air. About $\frac{3}{4}$ of all attempts at capturing grasshoppers were successful. On some occasions, a flying grasshopper that was being chased by a baboon was captured in flight by a lilac-breasted roller *(Coracias caudata)*. Yet these birds were threatened by the baboons only once (p. 194).

Locusts are grasshoppers that form immense migratory swarms after a long period of high rainfall. Locust swarms sometimes move through North and East Africa (e.g. during 1968). The impact of the baboons on the locust population—and vice versa—is not known, but because of the potential of biological control, the study of such ecological relations is of great importance. It may be, for example, that the devastating losses to farmers and ranchers that accompany an unchecked locust invasion exceed what is gained by eliminating baboons and other grasshopper predators from agricultural areas. KUMMER [1968] relates that the hamadryas that he watched ate large quantities of locusts after swarms of them settled in the area.

Three attempts (all in February) at catching flying butterflies (small, yellow-white) were observed. One attempt was unsuccessful; the outcome of the other two attempts is not known.

On a number of occasions, baboons were seen gnawing on pieces of rotten acacia wood or picking small objects off such wood with their lips or fingers. Perhaps they were eating termites, beetles or ants living therein.

No other invertebrates were observed being eaten by baboons in Amboseli. In Nairobi Park, we once observed baboons picking up and eating objects from a patch of soil on the upper bank of a stream. Subsequent examination of this earth showed it to be full of earthworm diggings and castings, and we assume that the baboons were eating earthworms.

One reptile, a small lizard, was caught and eaten by a juvenile-2 male of the Main Group. On another occasion, a large monitor lizard *(Varanus niloticus)* was closely observed by members of the Main Group but was not attacked by them.

Three-banded plovers *(Charadrius tricollaris)* sometimes made screaming dives on the baboons as the baboons foraged in open grassland, but predation on plovers or their eggs was not observed. One plover nest in which eggs were observed was empty a few days later; the predator is not known.

On April 21, two adult males of the Main Group fed on eggs of the helmeted quinea-fowl *(Numida mitrata)*. On May 15, a third adult male of the Main Group caught and ate two chicks of this species. As he ate them, a subadult and two juvenile males chased a covey of these chicks, but retreated when the adult fowl turned on them.

On May 14, a mouse or some other small mammal was eaten by an adult female of the Main Group.

The Cape hare *(Lepus capensis)* and an unidentified fossorial mammal were the vertebrates preyed upon most frequently by the baboons.

Of our 12 observations on baboon-hare relations, three were of hares that were within or near the Main Group but were not attacked by the baboons. At least one of these hares was definitely seen by the baboons. In 4 other cases, baboons of the Main Group were seen chasing hares. On 2 occasions, unsuccessful chases were made by an adult female. We observed only about 60 feet of a third chase; it is believed to have been unsuccessful. A fourth hare was at first chased by several members of the Main Group, then only by the most dominant male. He continued the chase for 70 sec., then caught the hare and ate it. In 2 cases in the Main Group, and 3 cases in other groups, we did not see the capture of a hare that was eaten. (In one of these cases an adult female in the Main Group already had a hare when we first saw it. The hare was taken from her by an old adult male, who ate it. In the other instance a Main Group juvenile-2 male had a hare when we first saw it; he ate it. Twice in other groups an adult male had a hare when the hare was first seen by us; both hares were eaten. We did not identify the predator baboon in the third instance.) Three additional observations of hares being eaten by baboons in our study area were made by Dr. THOMAS STRUH-SAKER; 2 were eaten by subadult males.

The fossorial mammal, a small, pinkish animal about three inches long and one inch diameter, was dug out of its burrows during the rainy season, when the earth was soft (observations November 24–April 29). In the Main Group, adults—male and female—and juvenile-2 males were observed digging for these animals, which were caught when they jumped or ran from their excavated burrows and were then eaten immediately. The burrows of these animals (mole-rats, Bathyergidae?) occurred in clusters, in bare ground near fever trees. We did not find anything in 14 holes into which we dug. In the 15th, we found a large scorpion. Yet in no case did we see baboons give the 'scorpion-response' that was described and illustrated by HALL [1963] when they dug into these burrows.

A neonatal Grant's gazelle *(Gazella granti)* was killed and eaten by adult males in HB Group.

Our attention was first caught by 3 adult gazelle that were running back and forth, lunging at the baboons. Within a minute, an adult male baboon appeared from a thicket with a baby gazelle in its mouth. The gazelle already had one wound other than where it was held by the baboon's jaws; it bleated. About 6 other adult males pressed close, but the male with the gazelle avoided them. During the next 20 min, there were repeated agonistic interactions among adult males around the meat (fig. 51). Several of the males became smeared with blood. Eating began on the posterior half of the body and progressed toward the head. Various other members of the group, including some adult males, watched this scene for a few minutes before moving on.

This gazelle was killed and eaten on November 21. Unfortunately, we were away from Amboseli December 4–February 8 (except for one day), and thus were not present during most of the time when neonatal gazelle were available. By the time of our return to Amboseli in February, the newborn gazelle were too swift-footed to be easy prey for baboons, and no further attempt to prey on them was seen.

Two other species of nonhuman primates live in the study area, the vervet monkey *(Cercopithecus aethiops)* and the lesser bushbaby *(Galago senegalensis)*. Both were eaten by baboons. Four vervets were chased, but escaped. One vervet was eaten primarily by a subadult male, one by three adult males, and two others each by a single adult male. In 6 cases (one in the Main Group, two others contributed by T. STRUHSAKER) baboons were observed eating vervets, but the capture was not observed.

In one other case, the entire predation process was observed. As the group progressed near fever trees, moving toward their sleeping grove, an adult male, Humprump, suddenly sprang at vervets in a group that had been displaced by the approaching baboons. He reached around a piece of dead wood hanging from a tree and with one swipe caught a vervet. Another adult male, Whitetip, followed him for a few minutes, but Humprump moved away from the group, into a clump of undergrowth. There, unmolested, he ate the vervet, beginning with the hind legs (fig. 52). The rest of the Main Group continued their progression. Forty minutes later, Humprump left the rest of the carcass and moved toward his group. He had eaten the hind legs but not the feet. He had completely crushed the mandible and skull case, and had apparently eaten the brain tissue. The arms, hands, thorax and viscera had not been eaten.

According to T. STRUHSAKER, the vervet was a female, 3–4 months old. Humprump is one of two males that migrated from the Main Group to the TTF group, where he was observed, some five months later, eating another vervet. While this may be coincidence, it seems more likely, in view of our infrequent observations on the TTF group, that male Humprump was particularly adept at catching vervets or was particularly attracted to them.

Bushbabies were twice observed being preyed upon by baboons. One bushbaby was eaten by an adult male, not of the Main Group

Fig. 51. Adult males of HB Group fighting over infant Grant's gazelle (held by male on left).
Fig. 52. Adult male Humprump eating a vervet monkey.

(capture not observed). Another bushbaby was chased in a dead tree, then caught and eaten by an adult male, whereupon other baboons began to tear bark from this same tree (observation by T. STRUHSAKER).

Once, as two duikers or dik-diks [48] chased each other through the undergrowth, beneath a grove of fever trees, they passed directly in front of the most dominant male baboon of the Main Group, who was sitting quietly on a log. The male jumped at them, but missed.

A small, unidentified mammal (?) that was caught and eaten by an adult female of the Main Group completes our records of baboon meat-eating.

6. Some Social Aspects of Feeding

The response of other baboons in a group when one of them had found a prey animal varied greatly. With very small prey, such as grasshoppers or the fossorial mammal, there was no competition. At the other extreme, there was much fighting among the adult males of HB group for the meat of the infant gazelle. While the prey animal was being chased, the baboons gave a chorus of barks that sounded just like the chorus given when the baboons themselves were faced with a predator.

Once the initial aggressive responses are over and one baboon has settled down to eat the prey, one or more other baboons of the group usually show a 'vulture response'. That is, they sit near the feeding baboon, quietly watching (fig. 53), and then, when the feeder walks away, pick up whatever scraps are left behind.

Baboons sometimes approach and sniff the muzzle of another member of their group, or at least, they place their muzzle up to the muzzle of the partner (fig. 54), and this behavior seemed much more frequent when the partner had food in its mouth. It may be that in this way, information about a new food source would spread through the group. Similar behavior has been described by HALL [1962, 1963] in chacma baboons and by STRUHSAKER [1967] for vervet monkeys (*Cercopithecus aethiops*), based on observations made in Amboseli.

Baboons sometimes supplanted one another from grass plants that had been partially dug up. Thus, by exploiting the labor of others, dominant baboons can get a larger supply of food with a smaller ex-

---

[48] Both have been recorded in Amboseli [WILLIAMS, 1967]; our brief view did not enable us to tell which we were seeing.

Fig. 53. Adult males of HB Group 'waiting for scraps' of the infant Grant's gazelle.

penditure of energy. As in the case of water (p. 138), strong displays of aggression during displacements from food were not noted. However, this point should be checked in subsequent field observations on baboons, because displacements during foraging were not noticed until fairly late in our study. The displacements usually were quite subtle, involving only simple approach-avoidance behavior over a distance of several feet between animals. The observer must have seen exactly which plant was being worked on by the supplanted individual. We have suggested elsewhere, in another context, that conservation of energy may be one of the major advantages of high dominance status [ALTMANN, 1967].

7. *Comparisons and Discussion*

(a) Water Resources of Other Baboons

KUMMER's 1968 monograph on hamadryas baboons in Ethiopia includes a number of observations on the use of water resources by these animals.

Fig. 54. 'Muzzle-muzzle' interactions.

'During the dry season the bands will usually come together at midday at some shady riverbed which still contains water.... Drinking is accompanied by a one to two hour rest period. Grooming and play occur at this time but chasing between males and copulation is as a rule omitted. In the rainy season, this midday rest is replaced by a number of short rest periods under bushes while the animals drink here and there as they cross streams and riverlets' [p. 12]. 'During the dry season, each troop had 2 to 4 permanent watering places within its range, mostly at pools under small chutes in the otherwise dry riverbeds. The hamadryas frequently dug individual drinking holes in the sand of the riverbeds, at any distance from their natural pools, by pulling up the wet sand with their hands while sitting. In contrast to the water of the pools, the water collecting in such holes is cool and free of algae. In order to drink, the baboons rest their weight on their forearms, often anchoring their tails around a piece of rock' [p. 164]. 'The thorny branches with which the [Gurgure] nomads protect their own water-holes were sufficient to keep the wild ungulates away from the water, but the baboons were seen dragging the branches apart and drinking from the holes' [p. 165]. 'In the dry season, it was striking to see that the large parties gradually assembling at a water hole did not drink at once but first settled down for grooming and play. On 6 occasions when the times were measured, an average of 69 min elapsed before the first baboons drank, the drinking being followed each time by another social period' [p. 169]. 'At the waterhole, around noon, there was a temporary and significant reduction in flight distance, and, as at the sleeping rock, the distance was significantly lower before the location-specific action than after its occurrence' [p. 170].

Digging of water-hollows has also been observed in anubis baboons, by CROOK and ALDRICH-BLAKE [1968, p. 214], on the sandy shore of Lake Langano, Ethiopia, 'the small holes rapidly filling with seepage from the lake'. The anubis that they observed near Debra Libanos, Ethiopia, 'drank directly from the stream or from puddles of rain water' [ibid.].

The anubis baboons that ROWELL [1966] studied along the Ishasha River in Queen Elizabeth National Park, Uganda, had ready access to the water of the river, which is permanent, fast-flowing, and cold. Furthermore, rain falls throughout the year in the area, and it is exceptional to have more than two weeks without rain. In addition, these animals often feed on succulant foods, such as figs. Although ROWELL's report does not include any data on drinking behavior, it seems unlikely that water is a limiting factor affecting the activities of this population at any time of the year.

Similarly, for the chacma baboons in the Cape Reserve, South Africa, drinking-water does not seem to be a problem. HALL [1962a] notes that 'in the 1958/59 observation period [one day of observation per week throughout the year with occasional extended periods up to 14 days], baboons of the Cape Reserve groups were seen to drink on only 8 occasions. Water is to be found at many different points in the Reserve during the rainy months of winter and spring, but it seems probable that the animals obtained most of their water from dew precipitation and from the water-content of the flora on which they fed. There was no evidence of regular drinking from dams or streams, and 7 of the 8 occasions occurred during the months of February, March and April 1959, when a few of the animals drank briefly from the small puddles of water in the hollows and crevices of rocks left by the occasional rain-shower. In the 1960 period [August and part of September], there were 8 occasions when a few baboons of S group drank from rain-puddles in the same way. They were not seen to drink from any other standing water in their area. The drinking posture has always been to crouch down so that the mouth comes into contact with the water. On 3 occasions, during this period, baboons of S group drank briefly from pools of water on the coast which were distinctly salty to taste. On one of these occasions, a fairly hot, sunny day, practically every baboon in this group drank from a salt-water pool close to the inter-tidal rocks, some drinking briefly, some for as much as thirty seconds' [pp. 215–216].

By way of contrast, the arid northern Transvaal is the site of the current work by STOLTZ and SAAYMAN [1969]. The baboons that they observed there drank every day, and the scarcity of water sources in the area brought groups together at drinking places. These authors have emphasized the importance of water 'in the determination of day ranges and in the utilization of home ranges.... The scarcity of water sources in this arid area...leads to an extensive overlap of the home ranges of troops near the water supply. Seasonal variations in the availability of water may bring about irregular movements of troops into unfamiliar terrain'. Their report states that, 'water is not constantly available in some reservoires and most natural drinking places'. As water sources dried up, the baboons shifted their day ranges to other areas [STOLTZ, personal communication].

HALL [1962 a] made observations on baboons in the arid thornveld of South-West Africa. In July, 1960, there was no rain or dew precipitation, and vegetation was very sparse. 'Groups were observed to drink at the infrequent water-holes used by game, at dams, and from cattle-troughs, on 10 occasions. Nine of these came in the period from 1400 to 1700 h, and the tenth occasion was a session from 0923 to 1020, when every individual of a group came down to a water-hole from the cliffs above and drank...' [p. 216]. Weather bureau data quoted by HALL indicate that in South-West Africa, '1959/60 was the fifth year of drought, the total rainfall over the 12-month period having been 123.5 mm', whereas in the Cape, 'winter rainfall reached a highest total figure of 118 mm in one month, but averaged about 30 mm a month. The lowest rainfall was 2.4 mm in December' [p. 189].

According to SHORTRIDGE [1934, quoted in HALL, 1962 a], baboons in South-West Africa '... drink regularly, and in undisturbed areas, usually come to water in the late afternoon just before sunset'.

HALL's observations on the chacma baboon's access to water in other areas have been summarized by him [HALL, 1963, p. 16] as follows:

'In S. Rhodesia, ... groups [49] came to the edge of pools to drink, or, on the Kariba island, drank from the lake waters in which, as also at Mana Pools, they paddled in the shallows in search of food, such as the stems of water lilies. During or after rain, the animals lick up water from rain-puddles and from their own fur.... Chacma baboons certainly have no aversion to immersing themselves in water. Thus, in the Cape... [a group of baboons] walked through the shallow water of a vlei [temporary pond or lake], some of them drinking, and while in the vlei, a severe squabble broke out in which several of the participants were drenched with water. This group, and the others that fed off shell-fish, frequently splash through sea-water pools, and confidently sit on rocks at the very edge of the waves.'

After subsequent field work on anubis baboons in Murchison Falls National Park, Uganda, HALL [1965 b] wrote that 'baboons drink fairly regularly from streams, rivers, or waterholes, and, indeed, the home ranges in all the groups observed in Murchison always included such water supply'.

HALL's comment, above, is one of the few descriptions of baboons entering fresh water. Another is by DEVORE and WASHBURN [1963], who write that '... along the rivers in Nairobi Park and around the water holes at Amboseli Reserve... they will wade into the shallow water to eat such plants as rushes and the buds of water lilies'. BROWN's book on Africa includes a photograph of baboons in a pool in a miambo woodland [BROWN, 1965]. We never saw baboons enter water during our study. It seems likely to us that avoidance of immersion, the digging of drinking holes, described above, and even the minimal contact between lips and water surfaces while drinking (fig. 46) are adaptations to minimizing infection with schistosomiasis. Schistosome infections in baboons have been reported by MILLER [1960] and NELSON [1960].

[49] HALL [1962 a, p. 183] comments that these may be cynocephalus baboons, not chacma.

DeVore and Washburn [1963] wrote:

'In 1959 throughout the dry season there were only two sources of water in Nairobi Park which also contained tall trees: the Athi River, which forms the southern boundary of the Park, and a water hole in the core area of the Lone Tree troop[50]. All the troops except Lone Tree had at least one core area along the Athi River, which is the boundary between the Park and the Ngong Reserve.'

(b) The Dry Season

We suggest the following explanation of the relative infrequency of drinking on the part of Amboseli baboons during the dry season, without discarding the possibility that a sampling error is involved. As the dry season progresses the temporary rain pools dry out until finally the only available sources of free water are the permanent waterholes. However, baboons are much more susceptible to attack in the vicinity of these waterholes—the surrounding vegetation makes an excellent hiding place for leopards—and drink from them only when necessary. Two lines of evidence support this explanation. The first is the distribution of predators and their attacks, which are clearly concentrated in the vicinity of the waterholes (see fig. 55, below). The second line of evidence comes from the relative frequency with which the baboons drank from permanent waterholes versus temporary rain pools. In the no-choice situation, i. e. at the end of the dry season, they drank only from the waterholes; but during the rainy season, when they had a choice, they seldom drank from the waterholes (table XXII).

Several mammals of the East African savannahs, including dik-dik, Grant's gazelle, gerenuk and oryx, can live indefinitely without access to drinking water [Lamprey, 1963, 1964; Taylor, 1968]. Impala gazelle are a marginal case in that they can survive with no drinking water other than morning dew [Lamprey, 1963]. Other mammals, including baboons, are limited in their distribution by the amount of time that they can go without water and by the distribution of water itself. Consequently, such species are restricted, at the end of the dry season, to the area within cruising range of permanent water sources.

[50] This description seems to differ from that given in DeVore and Hall [1965]: '... only the Athi River at the southern edge of the park contained surface water throughout the dry season.'

During the rainy season, most of these obligate drinkers disperse widely across the African savannahs. The baboons do not. Instead, they simply enlarge their home range. It may be that at such times, the distribution of baboons is restricted by some essential resource with a more restricted spatial distribution. The most likely limiting factor is the distribution of groves of trees that are of adequate size to provide night-time safety for the baboons.

Scarcity of water—or of any other essential resource—may have a marked influence on home range overlap. According to DeVore and Washburn [1963] baboons in Amboseli 'were tightly clustered around waterholes' at the end of the dry season, in September and October, 1959. 'Troops 51, 66, and 171 used only the northern pool [KB 3], and went north and west from the pools during the day (shaded area)'. The shaded area on their map 2 (which did not reproduce on the printed copies that we have examined) indicates that these three groups ranged within the northern portion of the KB forest and the savannahs north of there, almost to the Enkongo Narok swamp. No tight clustering of groups around waterholes was observed during our study, although there were times when groups were near each other at waterholes. Such aggregations were minor compared with the nightly aggregations of groups in the vicinity of sleeping trees.

According to Stoltz and Saayman [1969], 'the scarcity of water sources in the area frequently brought troops together at drinking places; the length of time spent by troop RB at the reservoires may partially be accounted for by interactions between troops converging on the same water supply.' These authors describe intergroup interactions in detail.

Despite the ecological significance of water for man and beast, accurate information on its distribution is difficult to find. A 1958–60 water supply investigation map of Amboseli (Kenya Ministry of Works, Hydraulic Head Office, map DRG No. H 20/11) shows the man-made boreholes and the large swamps, but none of the permanent waterholes in the KH or KB forests. Only two of these waterholes (probably KH-3 and 4) are shown on the most detailed map of Amboseli that is published by the Survey of Kenya (Series Y 731, edition 2-SK, 1:50,000). On the other hand, a young Masai *moran* who grew up in Amboseli was familiar with all of the permanent waterholes except KH-2, which is small and is hidden on all sides by vegetation. On the basis of this experience, it would appear that the most accurate information on water distribution can be obtained only by detailed ground surveys, supplemented by interviews with indigenous people.

Masai names for the Amboseli waterholes are as follows. KH-0 = Maji Moto ('hot water' in Ki-swahili), KH-1 = Naburru Waterhole, KH-3 = Ole Nkaiyia W. H., KH-3.5 = ?, KH-4a = Ole Kandu W. H., KH-5 = ?, KB-1 = Olchorro Losailagi W. H., KB-2 = Ilkitiruani W. H., KB-3 = Ole Partimo W. H., KB-4 = Purdul W. H., KB-5 = Ole Kununi W. H.

(c) Foods of Other Baboons

Several authors have listed species of wild plants that they have observed baboons to feed upon, e. g. ROWELL [1966], HALL [1962 a, 1963, 1965 a], DEVORE and WASHBURN [1963]. Quantitative data on relative use are, of course, much more difficult to obtain.

Baboons have a well-known predilection for agricultural products, and in some areas they feed extensively from farms. However, almost all of the agricultural products involved have been introduced into Africa within recent times, and hence the baboon's use of such food is of only marginal interest in attempts to understand the evolution and adaptive radiation of these animals, except as an indication of their great adaptability.

HALL [1963] summarized his numerous observations on the food of baboons in southern Africa as follows:

'The diet, in all areas of observation, has consisted mainly of a wide variety of plants, bulbs, seeds, leaves and stems. In the Cape, where our collection of food specimens has been most thorough, 94 plants have provided some part of leaf, root, etc., for food.'

According to HALL [ibid.], chacma baboons on the Cape Peninsula, South Africa, also feed on intertidal animals, including black mussels, limpets, sea-lice, crabs, sand-hoppers, and probably oysters, and there is circumstantial evidence that shells of the last are broken open with triangular-shaped stones. Other indigenous animal foods of chacma baboons include 'locusts, grasshoppers and ants, and the larvae— probably of termites—found in the hollows of dead branches' and perhaps scorpions, although HALL presents evidence that these animals elicit a startle response from some chacma baboons.

Anubis baboons that HALL [1965 a] observed in the Chobi area
of Murchison Falls National Park, Uganda, fed on the seedpods and
beans of tamarinds *(Tamarindus indica)*, the tough sausage-shaped
fruits of *Kigelia aethiopica*, occasionally the small fruits of *Balanites
aegyptica*, various grasses, mushrooms, and ants, and were reported
to eat bird eggs, and crocodile eggs dug up by monitor lizards on the
mud flats of the Victoria Nile. In none of HALL's field work did he see
predation by baboons on any mammal.

CROOK and ALDRICH-BLAKE [1968] observed anubis baboons in
Ethiopia, where baboons of this species feed on

'fleshy plants particularly the Prickly Pear, *Opuntia*, also *Phoenix* palm, Aloe,
bulbs dug up, olives both from below the trees and from the branches, fresh
twigs torn from trees (especially *Ficus* sp.), lichen obtained by gnawing at dead
wood, *Acacia* leaves and arthropods obtained by turning stones.'

These authors describe an unusual feeding technique:

'*P. a. doguera* feeding techniques may be highly elaborate. In the forest *Opuntia*
leaves and fruit were clearly the favoured food, large quantities of seeds being
seen in all faeces. This is a spiny plant and the animals had developed great skill
in handling both the fruit and the large plate-like fleshy leaves which were eaten
like sandwiches. The animals will sit gingerly on or near the *Opuntia* plant and
strike or stroke the chosen leaf from the stem attachment to the end of the
structure. This eventually breaks the point of attachment. The direction of strike
follows the lay of the spines so that the animal is not pricked. When the leaf
hangs loose, it is plucked and closely examined and fingered, all spines on it
being removed. Occasionally a spine sticks in the hand and is then cautiously
pulled out and discarded. Fruit may be picked, cleaned similarly and then bitten
into in such a way that the animal removes the centre of the fruit from the skin
which is then thrown away. Fruit is also rolled too and fro in a patch of dry
earth which evidently removes the prickles. Again the fruit is bitten into, the
contents eaten, and the skin discarded. Sequences such as these were observed
in detail more than 10 times and also filmed.'

In the Ishasha River valley of Uganda, 'a complete list of plants eaten
would probably be approximate to the botanical species list for the
area....,' according to ROWELL [1966].

'It was in relation to... [fruiting] trees in particular, that the baboons' local
knowledge seemed important, and each tree was exploited as it became edible.
Occasionally a baboon, usually an adult male, was seen to leave the main party
to inspect a tree, taste the fruit, and if it was not ready, to spit it out and return
to the troop.... There was no season at which a wide variety of food plants was

not available.... Many of the plant foods were protein-storing seeds, like *Parkia* and *Acacia*. Fruits were, for the most part, eaten long before they would be considered ripe by human standards.... Of grasses, the parts eaten were new shoots, green seedheads, and the storage leaf bases of some species, which are again high in protein content....'

'The mainly herbivorous diet was supplemented by animal protein. Insects (mainly grasshoppers and butterflies) were grabbed in the open grass; land snails *(Limicolaria)* were suspected but never confirmed as food (a baboon found and bit open a dead shell and expressed disgust when he found it full of mud); birds' nests were investigated and a passing bird grabbed at, but again eating was not seen. Occasional lizards were caught and eaten. On 6 occasions, hares *(Lepus capensis)* were coursed by baboons after being flushed in open grass, and on four of these they were successfully caught and eaten. This represents a catch rate of one every 30 h that baboons were observed in the appropriate vegetation, which over the year, would represent both considerable protein input into the population, and predation pressure on the hares. (One captor was a female, one a juvenile male, and two adult males.)'

KUMMER [1968, p.160] wrote of the hamadryas diet as follows: 'On certain days, Kurt recorded the food eaten by each baboon he could observe at the moment he spotted the animal.' Kurt's records indicate that beans and dry leaves of acacia accounted for 84% of the food types at the end of the dry season; acacia flowers and fresh shoots accounted for 46% of the food types in the sample during the long rains. The only other type of food that was observed being eaten by more than 10% of the individuals was grass seed, picked from the ground, which accounted for 44% of the food types sampled during the long rains. Such extensive use of acacias and of grasses is just what we observed in yellow baboons in Amboseli.

'Acacia flowers and grass seeds are preferred to any other type of food. Both are more abundant in the northern parts of the habitat. This is probably the reason why, during the rainy season, the daily routes pointed northwards from all rocks. ...Strikingly, the permanently green, large (more than 15 m) trees of the gallery forests were hardly used as sources of food.... The baboons only approached them when they went to drink and rest at the waterholes in the rivers, and then only a few animals would occasionally climb onto limbs not higher than about four meters and pluck some fruits' [ibid., p.161].

KUMMER continues: 'Though generally considered as extremely terrestrial primates, our "desert baboons" may be more arboreal in their feeding habits than the anubis baboons who sleep on trees but usually forage on the ground....'

DeVore and Washburn [1963] have written a general description of the feeding habits and foods of baboons, based primarily on their observations of anubis baboons in Nairobi Park, Kenya. According to them, 'grass is the baboon's single most important food'. These authors, too, indicate the important role of acacias in the baboon diet. Grasses, along with acacias and figs, are 'the most important food sources in the park'. Other plant foods indicated by these authors include tuberous roots or bulbs, rushes, buds of water lilies, the berries, buds, blossoms and seed pods of various bushes, flowering plants and shrubs, 'kei-apples', croton nuts, sisal plants, mushrooms, and the produce of native gardens.

Invertebrate food eaten by baboons in Nairobi Park included ants living in the galls of *Acacia drepanolobium*, beetles, slugs, crickets, and grasshoppers. During an infestation of the area with 'army worm' caterpillars, baboons ate little else. Some were timed picking up 100 caterpillars per minute.

DeVore and Washburn recorded several occasions during 12 months of study in Kenya and the Rhodesias on which they saw baboons eating freshly killed vertebrates, as follows: two half-grown hares *(Lepus capensis)* were eaten by adult males, two or three fledgling birds, probably crowned plovers *(Stephanibyx coronatus)*, were captured by a juvenile and eaten by an adult male, two very young Thomson's gazelle *(Gazella thomsonii)* were eaten by adult males, and a juvenile vervet monkey *(Cercopithecus aethiops)* was eaten by an adult male.

'In summary, baboons may be described as very inefficient predators...and meat never becomes an important source of food for the whole troop.... It would seem...reasonable to us, on the present evidence, to assume that meat has been a consistent but very minor part of the baboon diet throughout their evolutionary history. In localities where sources of animal protein can be obtained without danger baboons apparently include these in their regular diet.... But baboons are ill fitted anatomically to be carnivores, and too great a dependence on meat eating could have been detrimental to their wide exploitation of the vegetable foods they depend upon today. By their utilization of a wide variety of plant and tree products, baboons have been able to spread over the African continent, and, together with the macaques, to cover most of the tropical Old World' [ibid.].

These authors emphasize the frequency with which baboons are near potential prey animals and seem to ignore them. A female baboon near Victoria Falls caught a vervet and held it in her mouth, but then released it, unharmed.

Other observers have seen baboons eat vertebrate animals. The numerous observations of chacma baboon predation on domestic stock

in South Africa have been compiled by DART [1953, 1957, 1963] and by
STOLTZ and SAAYMAN [1969]. The most recent of DART's articles also
records several observations on baboons in game reserves preying on
wild animals, including mountain reedbuck, duiker, francolin, guinea-
fowl, steenbuck, and impala gazelle.

In the field work of STOLTZ and SAAYMAN [1969], involving 183
days, and 1,285 hours of systematic observation, only two instances of
meat-eating were seen. Once, a fully grown African hare was eaten, first
by a lactating female, then by an estrous female. On another occasion, a
young klipspringer *(Creotragus creotragus)* was killed by a dominant
male, then eaten by him, by another dominant male, and by a receptive
female, in turn. In both cases, other members of the group chased or
threatened the animal that first had the meat. Yet, '...hares and klip-
springers were seen moving without apparent fear close to baboons
which paid them not the slightest attention' [ibid.].

During KUMMER and KURT's work on hamadryas baboons, only
one case of possible predation on small mammals was observed: a
three-year-old hamadryas female was seen carrying a dead, young dik-
dik antelope *(Madogua kirki)*. When approached by an adult male, the
female fled into the bush, still carrying the dik-dik and screaming.
Subsequent examination of the carcass suggested that it had been
partially eaten [KUMMER, 1968].

BARTLETT and BARTLETT [1961] have observed anubis baboons
in the Serengeti (northwest Tanzania) kill Thomson's gazelle fawns
'not once but a dozen or more times'. According to them, 'the big male
baboons are the biggest offenders, searching slowly and systematically
through a herd of grazing Tommies or Grants [gazelles]. The baby buck
rely on camouflage for protection often not moving until they are
actually picked up from the ground.' The existence of frequent baboon
predation on Thomson's gazelle fawns in the Serengeti has been
confirmed by SCHALLER [personal communication].

LOVERIDGE [1923] recorded that a 'bushfowl' was taken from a
trap by a yellow baboon and eaten.

In recent years, several other species of nonhuman primates have
been observed preying upon vertebrates. Chimpanzees *(Pan satyrus)*
have been seen eating young bushpig and bushbuck, red colobus monkeys, baboons, and a small unidentified mammal [GOODALL, 1963;
VAN LAWICK-GOODALL, 1965]. Patas monkeys *(Erythrocebus patas)*
have been observed catching small lizards (including *Agama agama*) and

once eating the eggs of a tree-nesting bird [HALL, 1965 a, p. 50]. According to STRUHSAKER [1967], vervets in Amboseli ate the eggs of drongo birds *(Dicrurus adsimilis)* and chicks of the yellow-necked spurfowl *(Plermistia leucoscepus)*. Gibbons *(Hylobates lar)* eat bird eggs and young birds [CARPENTER, 1940]. At the Yerkes Field Station, where groups of primates are kept in large, outdoor enclosures, both pigtail macaques *(Macaca nemestrina)* and capuchin monkeys *(Cebus nigrivittatus)* catch and eat birds; the capuchins also catch and eat lizards [BERNSTEIN, personal communication]. Other cases of predation by primates on vertebrates, compiled from replies to a questionaire, have been listed by KORTLANDT and KOOIJ [1963].

Several aspects of baboon meat-eating deserve comment. First, no cases of carrion feeding were observed. Even when we did not observe the capture of the baboons' prey, the presence of fresh blood indicated that the animal had just been killed, and the baboons' antipathy toward predators makes it highly unlikely that they would ever take meat from them. We do not know how the baboons would respond to an animal that had recently died, say from disease. Such an animal would not remain on the ground for long, however: the numerous vultures of the savannah are ever-watchful.

Second, it is not always true that baboon predation is fortuitous, that they merely stumble upon their prey [DEVORE and WASHBURN, 1963] and do not hunt it. If one considers the major phases or components of hunting, including stalking or searching, attacking or chasing, killing, and eating, it becomes clear that the extent to which baboons 'hunt' depends upon numerous factors, including particularly the species of prey. The searching phase was clearly evident in visual scanning for grasshoppers, in digging out the fossorial mammal, and in tearing bark off the tree in which one bushbaby had already been caught. The lizard, the duiker, the guinea fowl chicks, and hares were attacked or chased. Stalking, in the sense of taking any action whose major function is hiding from potential prey, was never observed in baboons. Killing, including, when necessary, aggression against the prey animal's conspecifics (as in the case of the infant Grant's gazelle), was also observed, though the baboon that got most of the meat was not necessarily the one that did the killing.

Finally, the preponderance of males among the meat eaters is noteworthy, and such meat-eating may at times be a significant portion of their diet. During our study, the most dominant male of the Main

Group ate one hare, two guinea fowl chicks, probably several subterranean mammals, large numbers of grasshoppers, and just missed
catching a duiker. Male Humprump ate at least two vervet monkeys
within 5 months. Juvenile-2 male Three ate a small hare, several fossorial
mammals, and a lizard. Killing of young gazelle may predominate in
very large groups, such as the HB group, because of the capability of
such a group to cope effectively with the aggressive responses of a
group of adult gazelle, and the dominant males of such baboon groups
may thereby get a particularly large amount of meat.

(d) Food, Day-Journeys and Home Range Size

In describing the day-journeys of baboons, DEVORE and HALL [1965]
write:

> 'This varies from only a few yards (when a Kenya group sleeps in a fig tree and
> feeds in and under the tree throughout the following day) to a maximum distance
> of twelve mi. (observed once for a group of 65 in South-West Africa). The contrast
> in available foods between a heavily laden fig tree and the sparse vegetation of
> the study area in South-West Africa suggests that available food is the single
> most important factor affecting length of day range. This is supported by ob-
> servations at different seasons, which show that during the seasons when suitable
> vegetable foods are most plentiful average day ranges are shorter [HALL, 1962,
> p. 193]. A second reason for the longer average day range in the dry season
> (Kenya) or winter (Cape) is that a group is more likely to shift to a new core area...
> during these seasons. It is likely that this shifting is also related to the available
> food supply, representing movement to a new locus of foraging activity after
> reduction of the available food in the former locus.'

The available data on day-journeys (chapter V), though admittedly
sparse for some baboon populations, are consistent with the hypothesis
that the mean length of day-journeys is related to food abundance, and
thus may reflect the carrying capacity of the land. In arid areas, such as
South-West Africa, the northern Transvaal, and Ethiopia, plant productivity probably is limited primarily by available rainwater, and the
baboons must cover much ground in order to get enough food to sustain
themselves. At the other extreme, the gallery forest along the Ishasha
River, where ROWELL made her observations, is quite luxuriant; correspondingly, the groups moved much shorter distances each day. In
the Cape Reserve, which lies at 34° S. latitude and is thus well within the
temperate zone, seasonal variations in sunlight may play a significant

rôle in plant productivity, and, as HALL has suggested, this difference
in plant abundance may, in turn, account for the longer progressions
in winter.

This hypothesis does not contradict STOLTZ and SAAYMAN's
important correlation of maximum daily temperature with length of
day-journey (chapter V). Temperature may account for much of the
day-to-day variability within any one area, food availability may
account for long-term differences between areas.

Just as the distribution of food within the home range may affect
the movements of the animals, so the total available food in the area
may affect the total home range size. This has been suggested by HALL
on the basis of two groups that he studied in the Cape Reserve. One of
these groups (C group) had an area of 13 sq. mi., only a third of which was
extensively and regularly culled for food, the rest consisting of open
veldt with reeds, grasses, and some bulbs. 'In contrast, the home range
of S group covers an area of about 5 ½ to 6 sq. mi. but it is densely packed
with suitable vegetation, and only one small area at the northern end is
of the open veldt vegetation variety' [HALL, 1963]. The picture is
complicated, however, by the fact that group C, which had the larger
range, was about twice the size of group S.

(e) Selective Feeding

In reviewing our records of the baboons' food plants in the light of the
chemical composition of Kenya browse and pasture herbage, as analyzed
by DOUGALL, DRYSDALE and GLOVER, we have been impressed by the
apparent capacity of these animals to feed selectively on the most
nutritious parts of the plants available in their habitat at each time of
the year. Finding out how efficient they are at maximizing their yield
with a minimum expenditure of energy would be a fascinating research
problem, combining ecology, physiology, and nutritional biochemistry.

Corrigendum added in proof. Observations on baboons in Amboseli during 1969
indicate that a specimen of *Solanum incanum* was incorrectly collected instead of
Withania somnifera. The two grow near each other in moist depressions. The fruit of
the latter is bright red when ripe (September), and about the size of a small pea. Baboons
fed extensively on the ripe fruit of this plant. The fruit of *Solanum incanum* is not red
at any stage. When ripe, it is yellow and about 3–4 cm diameter. We do not know whether
baboons ever eat it.

VII. PREDATORS

1. Direct Responses

During 1006 h of observation on the Main Group, 103 reactions of the baboons to predators or supposed predators were observed. The diurnal and seasonal distribution of these reactions is shown in table XXIII. Their spatial distribution is shown in figure 55. Predator-reactions appeared to occur at the highest rate during the long dry season, next highest during the semi-dry period between the rains, and at the lowest rate during the intense rains of November-December. No diurnal pattern is apparent in the available data (table XXIII).

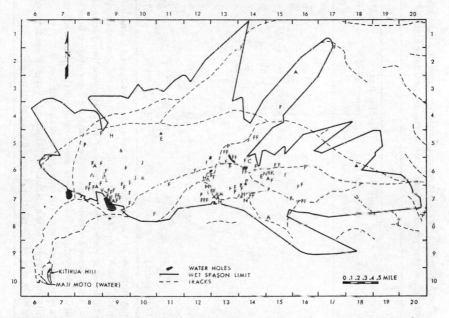

Fig. 55. Location of predator responses in the Main Group. *A* Alarm to unknown predator, *C* alarm to cheetah, *E* alarm to eagle, *F* false alarm, *G* intergroup alarm, *H* alarm to hyaena, *I* interspecific alarm, *J* alarm to jackal, *K* baboon killed by predator, *L* alarm to lion, *N* no response to presence of predator, *P* alarm to leopard, *S* secondary response, *V* alarm to vulture, ▲ leopard seen but no baboons nearby.

Table XXIII. Diurnal and seasonal changes in predator-responses by baboons of the Main Group. Reactions included are responses to leopard, lion, cheetah, minor predators, unknown predators, and false alarms

Time of day (beginning of interval)	Season									
	Jan-Feb (inter-rains)		Mar-Apr (long rains)		May-Oct (dry)		Nov-Dec (short rains)		Diurnal (all seasons)	
	Freq.	Prob.	Freq.	Prob.	Freq.	Prob.	Freq.	Prob.	Freq.	Prob.[1]
7	2	0.36	1	0.22			1	0.31	4	0.14
8			1	0.06	10	0.17			11	0.10
9	2	0.15	1	0.04	9	0.09			12	0.08
10	2	0.16	4	0.17	13	0.11	2	0.28	21	0.13
11	1	0.11	1	0.05	5	0.05			7	0.05
12	1	0.22	3	0.20	6	0.08			10	0.10
13			2	0.17	5	0.10			7	0.10
14					1	0.02			1	0.02
15					3	0.06			3	0.04
16			1	0.09	9	0.12			10	0.11
17					10	0.10			10	0.07
18					6	0.11	1	0.41	7	0.08
Seasonal (all hours)[2]	8	0.30	14	0.19	77	0.42	4	0.10	103	(0.10)

Freq. = frequency (number of cases). Prob. = probability, estimated from the relative number of cases per hour of observation (B+C time, table I) on the Main Group. (Separate tables for each type of predator response are available from the authors.) Each probability in the body of the table is the estimated likelihood of a predator reaction during the specified hour if a predator reaction occurs during the season indicated. Thus, these are estimates of the relative danger of each hour within the season.

[1] Each probability in this column is the estimated likelihood that if a predator reaction occurs, it will occur during the hour indicated. Thus, these are estimates of the relative danger at each hour.

[2] Each probability in this row is the estimated likelihood that if a predator reaction occurs, it will occur during the season indicated. Thus, these are estimates of the relative danger during each season.

The alarms and predator attacks were clearly concentrated in the forest areas and near waterholes (fig. 55). But this is to be expected, since these are the portions of the home range that the baboons enter most frequently and remain in longest. The relative danger of quadrats, in terms of the number of such predator-reactions per minute of observed quadrat occupancy and per quadrat entry are given in table XXIV.

Table XXIV. Predator reactions in quadrats. The quadrats have been rank-ordered on the basis of reactions per entry of the Main Group into the quadrat

Rank	Quadrat	Frequency of reaction	Reactions per entry	Cumulative reactions per entry	Reactions per minute
1	1,14	1	1.00	1.00	**0.028**
2	1,17	1	1.00	2.00	**0.012**
3	4,15	1	1.00	3.00	0.008
4	**7,9**	14	0.82	3.82	0.004
5	**7,8**	4	0.57	4.39	0.003
6	2,16	1	0.50	4.89	**0.022**
7	7,16	4	0.50	5.39	0.008
8	**6,14**	6	0.37	5.76	0.003
9	8,10	1	0.25	6.01	**0.016**
10	**6,8**	7	0.22	6.23	0.004
11	**5,14**	3	0.21	6.44	0.003
12	**7,12**	7	0.21	6.65	0.002
13	**7,14**	7	0.21	6.86	0.003
14	**6,9**	6	0.19	7.05	0.005
15	**6,15**	5	0.19	7.24	0.001
16	**7,13**	8	0.18	7.42	0.001
17	**6,10**	3	0.16	7.58	0.002
18	**5,13**	3	0.15	7.73	0.002
19	8,15	1	0.14	7.87	0.004
20	**6,13**	5	0.13	8.00	0.001
21	**6,12**	6	0.12	8.12	0.001
22	5,8	2	0.12	8.24	0.003
23	5,9	2	0.11	8.35	0.002
24	6,16	1	0.11	8.46	0.001
25	5,11	1	0.10	8.56	0.001
26	7,15	2	0.08	8.64	0.002
27	6,11	1	0.05	8.69	0.001

The 14 'favorite' quadrats (table XIX) are indicated in bold-face in column two. The four most dangerous quadrats, in terms of reactions per minute, are indicated by bold-face in column 6.

Some of these relations are shown graphically in figures 56 and 57. Figure 57 indicates that, for quadrats in which we observed the Main Group for at least 10 h, the number of alarms per quadrat is roughly proportional to the amount of time that the group spent in the quadrat. On the other hand, the frequency of predator-reactions per quadrat was poorly correlated with the number of times the quadrat was entered (fig. 56).

The Main Group's 14 'favorite' quadrats, which accounted for 74.8 % of the baboons' daytime (table XIX) also accounted for 81.5 % of their predator reactions (table XXIV). Another way of looking at this relationship is to observe that there were 0.00245 (= 84/34270) reactions per minute of observation in the 14 favorite quadrats, and 0.00136 (= 19/13920) reactions per minute in the rest of the home range. Thus, it appears that, as a group, these 14 quadrats are more dangerous than the others and that the high frequency of predator attacks in these 14 quadrats is higher than would be expected, considering the amount of time that the baboons spent in these areas; $P(z \geqslant 2.34) < 0.05$.

From the standpoint of minimizing exposure to predation, the baboons would be better off concentrating their activities in low-risk

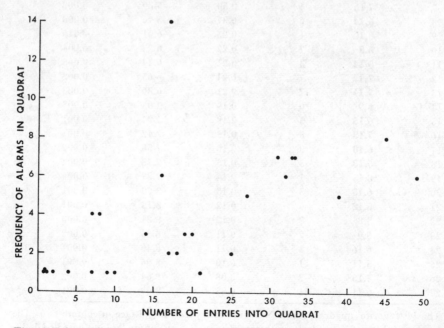

Fig. 56. Relationship between number of entries into quadrat and frequency of alarms in the quadrat.

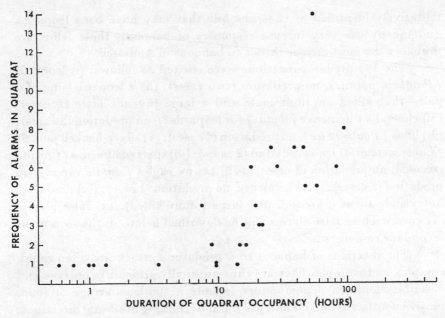

Fig. 57. Relationship between the duration of quadrat occupancy and the frequency of alarms therein.

areas. That they do not is attributable to the fact—documented in chapter IV—that a number of essential resources are concentrated in the favored areas.

Actually, none of these 'favorite' 14 quadrats are among the 4 quadrats that would seem to be most dangerous, as judged by the number of predator-reactions per minute of observation (table XXIV). In each of those, there were more than 0.01 predator-reactions per minute, with an average of 0.018 reactions per minute. These 4 quadrats accounted for 3.8% of the predator-reactions, but only 0.4% of the baboons' daytime positions (table XIX). All 4 of these quadrats are on the periphery of the group's range, and the high frequency of predator-reactions, many of which were false alarms (see below), may be primarily an indication of the greater tension of the group when in such areas, rather than of any real danger. In general, actual predator attacks (fig. 55) occur so seldom that it is quite difficult to use them to obtain a direct estimate of the relative danger of a quadrat. Wooded areas were the locales of most of those attacks that we did observe, and of all those that involved either leopards or animals that were seen briefly and tentatively identified as cheetahs but that may have been leopards. Judging by the very intense responses of baboons, these felids are probably the most serious threat to baboons in Amboseli.

The 103 predator-reactions were elicited as follows: (i) leopards *(Panthera pardus)*, no predation (two cases); (ii) a leopard—one of a pair—that killed an adult male and a large juvenile male (1 case); (iii) cheetahs *(Acinonyx jubatus)* (or leopards ?), no predation (2 cases); (iv) lions *(Panthera leo)*, no predation (4 cases); (v) silver-backed jackals *(Canis mesomelas)*, no predation (2 cases); (vi) spotted hyaenas *(Crocuta crocuta)*, no predation (3 cases); (vii) tawny eagles *(Aquila rapax)*, no predation (2 cases); (viii) a vulture, no predation (1 case); (ix) predators not visible to us (15 cases, one large infant killed); (x) false alarms (71 cases). These false alarms will be described below, in the section on secondary responses.

The reactions of baboons to a predator's attack are often rapid, complex and variable. They are thus unusually difficult to observe and describe. Common denominators include a sudden barrage of loud, plosive vocalizations, and flurry of animals running in several directions. Thus, if a leopard misses in his first lunge, the ensuing commotion probably makes it very difficult for him to single out one individual for attack.

When the baboons took flight, there was a tendency for some adult males to start later, to run more slowly, and to stop and turn sooner, the net result of which was that adult males tended to be interposed between potential sources of danger and the rest of the group. However, in some cases, adult and subadult males, and even juveniles and adult females with infants clinging to them lunged at or chased a predator.

When two silver-backed jackals chased juveniles and adult females of the Main Group, they were, in turn, vigorously chased and routed from the group by the most dominant male. On another occasion, two jackals approached an adult female, Shorty, on the edge of the group; Shorty was carrying an infant on her belly at the time. She fled into the group. A large young juvenile baboon then trotted toward the jackals, but ran off when the jackals turned and faced him. The jackals then left the area.

Once, an animal quickly identified as a cheetah, but more likely a leopard, sprang from undergrowth into the group, at the edge of the basin of waterhole KB-1. The baboons sprang away, then turned on the leopard, barking loudly as several members of the group ran at the leopard. The vocalizations were recorded. While details are difficult to note while recording, we did notice that at one moment, the dominant male was closest to the leopard. Faced with this mass attack, the leopard turned and ran into the catchment basin of the waterhole. Afterwards, we noted several wounds on the baboons. Adult male Whitetip had blood on his mouth. The subadult male had cuts on the midcallosity and the left hip. A juvenile-2 male had a long cut down the right forearm and on the right leg. He began to limp about 3 weeks later, favoring the right arm, but later recovered.

Another time when a 'cheetah' (leopard ?) jumped at a vervet on an *Azima* bush, adult male and female baboons, some with infants clinging to their bellies, ran at the cat, barking loudly. The cat quickly vanished.

On one occasion, while the baboons of the Main Group were fighting over the meat of an African hare *(Lepus capensis)*, a tawny eagle *(Aquila rapax)* swooped low over the baboons. The dominant male of the group, who had killed the hare, jumped up into the air at the eagle, but missed. The eagle landed in a nearby dead tree, where it was harrassed by juvenile baboons, who climbed into the tree and threatened it. (Cine films were taken during this encounter.)

Baboons do not invariably attack or approach a predator that has attacked them.

When a leopard sprang from the bushes into the midst of the Main Group, at the east end of a waterhole (KB-5), the baboons fled, barking, from the basin of the waterhole. (Sound recordings were made throughout this interaction.) The leopard looked down and very rapidly scanned the grass around him, where the baboons had been, then ran after the group for about 30 ft. before stopping. Then a baboon that had been crouching silently in the grass sprang up and ran to the group. Apparently, the leopard had overlooked it.

The leopard's scanning behavior suggests that this silent crouching is not an unusual response of some baboons. We saw it again, on May 7, 1964, in response to a strong earthquake that was felt throughout much of East Africa.

Most of the members of the group fled, then stopped about 80 ft. away, but a large infant froze in the grass. A small juvenile that was in a tree at the onset of the earthquake remained there. About a minute later, the infant got up and ran to the group. It ran immediately to the nearest baboon, an adult male, and rode ventrally on the male. After about 4 min the juvenile descended the tree and ran toward the group. As it did so, an adult female went toward it and reached out to it. But it ran past her and began rough-and-tumble play with another juvenile.

Baboons bark vigorously at a leopard whenever they see one, even if the leopard does not attack. But we once saw a large group of baboons virtually ignore an adult male lion who was basking in the afternoon sun. He was barked at by only one baboon, a juvenile-1 male, and by a group of vervets, while the rest of the baboons continued to move toward their sleeping trees, some passing within about 100 ft. of the lion with no indication of alarm. Yet, lions are known to prey on baboons, or at

least, to eat them (see below), and baboons sometimes react quite strongly to them (p. 79).

The responses of the baboons to Masai tribesmen, or to their dogs, was unlike their response to any predator. Even if they were in the trees when the Masai approached, the baboons fled across the ground, *en masse* and without vocalizing. (Occasionally, a single alarm bark was given when the Masai first appeared.) Although Masai are basically herdsmen, rather than hunters, we have observed a Masai boy using baboons as a target for spear practice.

2. Deaths from Predation

In addition to 3 deaths in the Main Group due to predators, one female of the Main Group, female Notch, was barking and 'looking for' her infant on the morning of July 11. She had long scratches down her left leg, and she limped. Her infant had been present during our previous observation period, on July 6, but was never seen again. We assume that the infant was taken by a predator, perhaps a felid, and that Notch was scratched while attempting to protect it.

We observed attacks upon the baboons of the Main Group by leopard, tawny eagle, and silver-backed jackal. In addition, we found a lioness eating an adult male baboon, and a striped hyaena *(Hyaena hyaena)* carrying half an adult male baboon in its mouth; neither of these baboons was from the Main Group. Hunting dogs *(Lycaon pictus)* were observed in Amboseli only once, and then, not near baboons.

3. Responses to Dead Infants

As has been recorded numerous times in the literature on primates, adult females sometimes carry their dead infants for several days, often until the corpse is dried and shriveled. We observed this in several instances during our study.

Two deaths during our study elicited unusual responses.

On May 15, the infant of Female Three was apparently killed by a predator. Most of the members of the group had left a grove of fever trees and had moved south, out onto open grassland, where they foraged. At 1155, one baboon of the group, probably a juvenile, began barking from a tree to the north, in the area that the group had just left.

From another tree, 100 yards away, an adult male of the group gave a deep, bellowing call. We then counted baboons in the group: all except one—the infant of Female Three —were accounted for. When the barking began, about half the baboons of the group looked that way, then returned to feeding. Only two baboons, females Three and Calico, responded differently: they walked back about 50 ft., then stood bipedally, looking toward the trees.

At 1200, when the barking ceased, we stopped observations in order to catch up on field notes. But at 1208, there was a sudden series of alarm barks from the north. We drove to that area. When we arrived, Female Three was walking rapidly and steadily away. Male Five, the dominant male of the group, had the infant of Female Three, and female Calico who may have been the infant's grandmother[51], was attempting to get the infant from him. He repeatedly threatened her, however, until finally she lay down on her side, exhausted. About 5 min later, she, too, walked away, leaving the infant with Male Five.

For the next 10 minutes, the adult male repeatedly rubbed the thorax of the infant. Several times, he took it into his mouth. Undoubtedly, he got a taste of the blood that was on the fur.

These manipulations by the adult male were strikingly like those used by baboons, including this male, when eating a piece of meat. The main differences were that the rubbing was very much prolonged, and the male did not eat the infant, he simply walked away. If the infant had been in a state of shock, rather than dead, the thoracic massage might have revived it[52].

Another infant, that of Shorty, was apparently found dead by a juvenile-1 female, perhaps its older sibling. When we caught sight of her, she was dragging the dead body, using one arm to press it to her ventrum. She carried it up to an adult female and grimaced at her. But this female ignored the infant and moved on. Shorty, however, accepted the dead infant and held it in much the same manner that a live infant would be held. The juvenile female then sat down and groomed Shorty (fig. 58). Thereafter, Shorty carried the dead infant about from place to place. Sometimes, when she put it down for a while, it was examined and even carried by other members of the group.

On the afternoon of the third day after the death of the infant, Shorty abandoned it. Several times that morning, as she foraged with the group, she lagged behind, then turned back and retrieved the corpse. By noon, she had begun to give the two-phased 'alarm' bark several times, looking back toward where she had left the infant, before going back. However, the other baboons did not become alarmed.

Beginning at 1331, she began a long session of barking. We counted 124 barks by Shorty during the next 89 min. Several times during this period, she moved short

[51] Female Calico was old enough to be the grandmother of the dead infant.
[52] This remarkable behavior of Male Five was filmed. An excerpt is included in the BBC-TV film on baboons 'Not So Much a Species, More a Way of Life'.

Fig. 58. Adult female Shorty holding her dead (or dying) infant and being groomed by the juvenile-1 female that brought the infant to her.

distances in the direction of the group, then at 1405, she climbed 8 ft. into a dead tree, faced in the direction from which the group had come and toward the area in which she had left the infant, then continued her barking. One adult male, Newman, remained with her. The other members of the group continued their progression, and none of them seemed to react to Shorty, except perhaps to delay somewhat their departure from the area.

Fifty-five minutes later, Shorty stopped barking, descended to the ground, gave one more bark, then trotted toward the group, which was 0.3 mi. ahead of her. She paused after 30 yards, looked back once, then again trotted toward the group. She gave no more barks until that evening, when she briefly stood bipedally, looked toward the distant grove in which the dead infant had been left, and barked.

4. Secondary Responses

In addition to responses to the immediate presence of a predator or presumed predator, such as we have described, there are other, more subtle responses of baboons that are indicative of the long-term influence of predators. These secondary responses are described below.

(a) Vigilance Behavior

There is a spectrum of responses, including alerting and visual scanning, which may be referred to as *vigilance behavior*. These responses are given quadrupedally, bipedally, or from an elevation, such as a log or tree. Movement of baboons into trees is common when they approach an area with obscuring undergrowth. Such animals are not 'posted sentries', despite several anecdotal accounts that so claim. HALL [1960], STOLTZ and SAAYMAN [1969] and BERT et al. [1968 b] have reviewed the literature on vigilance behavior, in both baboons and other animals, and have presented their own observations on such behavior in baboons (see also p. 137, above). From HALL's account, it is clear that the baboons in the Cape of Good Hope Nature Reserve, where he worked, were far more sensitive to the presence of a human than the baboons that we observed in Amboseli.

(b) Tension and False Alarms

Increased muscular tonus, quick 'startle' responses, alertness and frequent visual scanning are indicative of increased tension and lowered response thresholds. When in this condition, baboons became more cohesive, and laggards in the group progression quickly moved toward the group. Such tension and cohesiveness, combined with vigilance behavior, were pronounced whenever a group of baboons came to a narrow pass in obscuring foliage. The animals coalesced at the entrance to the pass. Some climbed trees and scanned the area. Often, there were several incipient progressions of clusters of animals before the group as a whole moved through. When they finally did so, many of them trotted or ran.

False alarms were likely to be given at such times, but they were given at other times, as well. False alarms included reactions that were similar to those given to predators, but in situations in which we were confident that no predator was present. The baboons' false-alarm response commonly followed a simple sequence: run → stop → look back → calm down. At times, some of the animals ran toward and even partly up trees.

This chain of reactions was sometimes precipitated by an alarm bark from just one member of the group. At other times, the animals did not vocalize at all, and the response apparently spread as a result of the baboons' sensitivity to any sudden movement in their group. In some cases we could see what elicited such false alarms. Examples are: (i) juveniles startled and ran when an adult male jumped at a flying insect; (ii) about 6 members of the Main Group ran about 10 ft. when a large juvenile suddenly jumped up on a log; (iii) a subadult male gave alarm barks to the broken stump of a large fever tree branch that was yellow with blackish spots; (iv) sudden rustling of dense, nearby undergrowth, either by baboons or by other animals, precipitated a strong alarm response.

We occasionally saw a baboon put its tail upright as it walked through dense foliage, in a posture used by baboons that have been frightened during social interactions. But none of the other components of the fear response—grimacing, screeching, cackling, piloerection, or flexion of the arms and legs—were present, and we have no explanation for this response.

(c) Short-Term Shifts

As a result of recent experience with a predator, the baboons may avoid a particular area, or may become hypersensitive to danger signals in an area. We arrived at the sleeping grove of the Main Group at 0930 one morning, to find a leopard in the trees with them, standing over two bodies. A second leopard was sitting on the ground, near the base of the trees. The baboons were still in the trees, barking loudly. As mentioned earlier (p. 80), the Main Group was never again seen sleeping in this grove. Three days after that incident the baboons were sufficiently alarmed by the chatter-bark of vervet monkeys in the KB woods that they left, moved to the KH woods, and slept there.

This last observation led us to wonder whether predator-alarms or false alarms, either in the group or in other, nearby groups, might have some general influence on the choice between sleeping in the KH woods or the KB woods. This possibility was tested as follows.

Those days on which we observed the Main Group and for which both the morning and evening sleeping grove of the group is known, were classified two ways, first, according to whether the baboons slept in the same woods (KH or KB) both nights or moved to the other woods, and second, according to whether there was at least one observed predator-reaction that day in the woods that the group slept in that morning. For the purpose of this tabulation, we included as a predator-reaction not only direct responses and false alarms in the Main Group, but also, conspicuous alarm responses in other groups, of the sort that the Main Group probably perceived and by which they might have been influenced. The results are shown below:

	Alarm	No alarm
Move	10	8
Stay	33	34

A test for independence in this two-by-two table gave $P(\chi^2 > 0.1110) > 0.05$. Thus, the available data do not reveal any effect of predator-reactions on the choice of woods in which to sleep.

Progressions sometimes stopped and were deflected when the baboons heard alarm calls ahead of them, or even when they began to approach an area in which they had previously encountered a predator. Similarly, if the group had been stationary, incipient progressions by subgroups toward such areas sometimes aborted, with the group finally going off in a different direction. Progressions away from areas where predators were encountered or alarm calls were heard were often prolonged and rapid. For example, during the first 30 min after descending from the trees in which the leopard had killed two baboons, the Main Group moved 0.61 mi., the longest we observed them to move in such a period of time.

(d) Long-Term Shifts

Persistent patterns of habitat preference and avoidance may result in part from repeated experience with predators in particular habitat types. For example, the great wariness of baboons when moving through an area with dense, obscuring undergrowth, such as *Sporobolus robustus* grass, may be due in part to their previous experiences with leopards in such areas. Not surprisingly, baboons tend to avoid such areas, and to prefer more open portions of their home range. It may also be that the extensive use of rain pools, rather than waterholes, and the preference for certain sleeping groves is the result of such experience.

On the basis of limited observations, we got the impression that each baboon slept as far out on tree limbs as it could, considering its

weight. This, too, might be the result of previous experience with
predators, as, indeed, might be the basic baboon pattern of sleeping in
trees or on cliffs. Now, however, we have begun talking about experiences
of baboons that may be consistent over many generations, thus selecting
for the genotypes of individuals who make the appropriate response.
Whether the differences between those that respond appropriately and
those that do not are due to differences in genetics, in social learning and
cultural transmission within the group, or in the direct experiences of
the individuals, cannot be decided on the basis of evidence now available.

5. Indirect Responses

A group of baboons often responds to the alarms or even alerting
responses of other baboon groups or other species. In response to the
alarm calls of another group of baboons, the Main Group gave any of a
variety of responses, depending upon circumstances. If the other group
was distant or the alarms were mild, the group generally alerted. Often
some individuals would stand bipedally, or climb trees, and look. If the
group had been moving toward the area from which the call came, they
sometimes changed the direction of their progression. In response to
rapid, intense alarms from another group at short range—say, within a
quarter mile—the group almost invariably joined in the barking.

Responses of baboons to the alarm calls of other species will be
described in the next chapter, on interspecific relations of baboons.

6. Comparisons and Discussion

(a) The Baboons' Predators

HADDOW [1952], in an excellent review of the predators of African
monkeys, writes: 'Leopards are frequently mentioned as enemies of
monkeys generally. There are numerous records of baboons being attacked and killed, and the writer has had personal experience of such
occurrences.' In a study of predation by lion, leopard, cheetah, and wild
dog in the Serengeti area, KRUUK and TURNER [1967] reported only
the leopard as a predator of baboons, which made up 4% of leopard
kills. In a study of predation in the Kafue National Park, Zambia,

baboons were the prey of leopards twice, out of 96 kills, but there were no instances of them being preyed on by lion, cheetah, wild dog, crocodile, or martial eagle [MITCHELL, SHENTON and UYS, 1965].

Baboons were not included in the stomach contents of 201 black-backed jackals *(Canis mesomelas)* from southern Africa, mostly from the Transvaal [GRAFTON, 1965]. STOLTZ and SAAYMAN [1969] write that jackals of this species 'were seen on occasion to move freely amongst baboons'. This contrasts with the vigorous responses of baboons in Amboseli to the approach of silver-backed jackals (p.176). GOODALL [1965] reported once seeing a chimpanzee eating a baboon and once observed chimpanzees unsuccessfully stalking a baboon.

Eagles (species not indicated) are reported to capture young baboons in the Drakensberg Mountains (see below). However, there appear to be no published observations of predation on baboons by the crowned eagle or 'monkey-eating eagle' *(Stephanaoaetus coronatus)*, an immense bird with a pan-African distribution, although these birds are known to prey on some monkeys [HADDOW, 1952, and references cited therein].

On occasion, pythons *(Python sebae)* may prey on baboons. ISEMONGER [1962] observed pythons attacking baboons on two occasions, once successfully, and he once observed baboons fighting with a python at night. On the other hand, PITMAN [1938, cited in HADDOW, 1952] does not include monkeys in his extensive list of the mammals preyed on by *P.sebae*, and he stated in a personal communication to HADDOW [ibid.] that neither he, nor LOVERIDGE, nor various other experts on snakes with whom he had communicated knew of monkeys or baboons being preyed on by pythons. ISEMONGER [ibid.] also reports seeing a young baboon bitten by a puff-adder.

Attacks of dogs on baboons have been reported by DEVORE and HALL [1965a], by CROOK and ALDRICH-BLAKE [1968], and by STOLTZ and SAAYMAN [1969].

Combining the observations in the literature with our own, we see that baboons have been killed and/or eaten by lion, leopard, striped hyaena, chimpanzees, python, puff-adder, 'eagles' and humans. Unsuccessful or incipient attacks on baboons have been made by leopard, cheetah, python, chimpanzees, and dog. In addition, baboons react strongly to the sight of leopard, and sometimes to lion, cheetah, spotted hyaena, and dog.

There is no doubt that within recent times, humans have been the major predators of baboons. Each year, thousands are killed or trapped, not only near farms, but also elsewhere.

The impact of the other predators on baboon populations cannot be assessed accurately on the basis of data now available. To do so will require long-term observations both on the baboons and on the predators. Perhaps, as a result of the present surge of research on African animals, the necessary information will gradually be amassed.

(b) Reactions to Predators

Several authors have described the way in which baboons approach and bark at cheetah, leopard, and lion [e.g. BOLWIG, 1959; FITZPATRICK, 1907; STEVENSON-HAMILTON, 1947; LOVERIDGE, 1923]. According to DEVORE and WASHBURN [1963],

'If a predator is sighted, all the adult males actively defend the troop. On one occasion we saw two dogs run up behind a troop, barking. The females and juveniles hurried ahead, but the males continued walking slowly. After a moment an irregular group of some 20 adult males was between the dogs and the rest of the troop. When a male turned on the dogs, they ran off. On another day we saw three cheetahs approach a troop of baboons. A single adult male stepped toward the cheetahs, gave a loud, defiant bark, and displayed his canine teeth; the cheetahs trotted away.... If baboons come upon predators while en route to their sleeping trees, the troop stops and waits while the males in the center move ahead and find an alternate route (the young juveniles and mothers with infants stay behind with the peripheral adult males). Eventually the dominant males return, the original order of progression is re-established, and the troop proceeds along the new route. These behavior patterns assure that the females and young are protected in the troop's center.'

Similarly, STOLTZ and SAAYMAN [1969] write:

'When packs of dogs threatened troops...the large adult males immediately interposed between the troop and the attacking animals and it was not uncommon for a single dominant male to maim or kill three or four large dogs before retreating in the direction taken by the troop.'

On the other hand, ROWELL [1966] says that,

'this deployment—fleeing troops forming a bodyguard of adult males between the source of danger and the rest of the troop—was seen on occasion, but only when the cause of alarm was so slight that the more confident adult males did not respond to something that set the juveniles running; a stronger stimulus produced precipitate flight, with the big males well to the front and the last animals usually the females carrying heavier babies' [ROWELL, 1966, p.362, rearranged].

Such precipitous flight, with the dominant adult males well ahead of the weaker animals and mothers carrying infants was observed by STOLTZ and SAAYMAN when chacma baboons were threatened by humans.

HALL [1963] wrote that

'ne attacks have ever been witnessed on a group by the writer. On one occasion only, the writer has seen most of a baboon group, in immediate response to a yap-bark, shrieking and scattering up the trees (Mana Pool, 1961), on the approach of a hyena which, in the early morning soon after the group had descended from their sleeping trees, was seen to be walking toward them.... The only other animal (non-human) reported to prey on the Chacma is the Black eagle, *Aquila verreauxi*. In the Cape Reserve, where a pair of these birds breeds and forages regularly, the baboons ignore their presence overhead, and no attack or swoops by the eagles toward the baboons have ever been observed. In the Drakensberg Mountains of Natal, Mr. BILL BARNES has reported himself seeing eagles take young baboons and quite often swooping over them.'

When HALL [ibid.] placed a stuffed serval cat near baboons in the Cape Reserve, it elicited little more than close attention, tension-yawning, and displacement mounting. No alarm barks were given.

ROWELL [1966] wrote that 'Baboons alerted and made alarm noises when non-hunting lions walked near the troop in the early morning on two occasions. Alarm noises of baboons, "singing" of colobus and sawing of leopard were heard together in the night on two or three occasions. No other interactions with these predators were seen or deduced.'

According to STOLTZ and SAAYMAN [1969], 'perseverative barking of the dominant males was heard on occasions from troops W and RB as late as 2400 to 0200 h; following such disturbances, the spoor of leopards or hyena were not infrequently found beneath the sleeping cliffs next morning.' During daytime observations, STOLTZ and SAAYMAN found that females with enlarged sexual skin were more likely than other females to give alarm barks.

(c) Order of Progression

DeVore and Washburn [1963] believe that they discerned a particular formation in a moving group of baboons, and that this formation is an adaptation to protection of the group:

> 'A baboon troop that is in or under trees seems to have no particular organization, but when the troop moves out onto the open plains a clear order of progression appears. Out in front of the troop move the boldest troop members—the less dominant adult males and the older juvenile males.... Following them are other members of the troop's periphery, pregnant and estrus adult females and juveniles. Next, in the center, comes the nucleus of dominant adult males, females with infants, and young juveniles. The rear of the troop is a mirror image of its front, with adults and older juveniles following the nucleus and more adult males at the end. This order of progression is invariably followed when the troop is moving rapidly from one feeding area to another during the day, and to its sleeping trees at dusk.... The arrangement of the troop members when they are moving insures maximum protection for the infants and juveniles in the center of the troop. An approaching predator would first encounter the adult males on the troop's periphery, and then the adult males in the center, before it could reach defenseless troop members in the center.'

Our observations do not confirm this description. We have numerous records of group progressions, not only in the Main Group, but in various other groups, both in Amboseli and elsewhere. The progressions that we observed, even those of a single group, did not reveal any invariable order of progression, nor was it true that the front and back of the progression invariably consisted of adult males and older juvenile males. Similarly, Stoltz and Saayman [1969] write:

> 'Clearly defined formations of this nature were not observed in the present study. Under these conditions, where the majority of movements occurred when the troop foraged widely scattered through the bush, two observers, keeping in close radio and visual contact, were unable to formulate meaningful conclusions concerning the spatial relationships of the various classes of baboon with reference to the dominant and... [subordinate] males. There was a tendency, however, for mothers with infants to remain in the vicinity of the dominant males and direct observation of troops as well as spoor counts of troops crossing open ground supported this impression.'

The actual variability of progression order in baboons when on the move deserves careful study, particularly in view of DeVore and Washburn's interesting suggestion that the formation assumed by a moving group of baboons is of great significance in the baboons' adaptations to a terrestrial mode of life. Our data on this topic will be analyzed and presented in a future publication.

VIII. OTHER ANIMALS

In the two preceding chapters, we have described the reactions of baboons to their predators and their prey. From time to time, baboons interact with other animals, and some of these interactions are mutually beneficial. Several times, we watched zebra, wildebeest and baboons move as a group to a waterhole. Associations of baboons with several species of ungulates at waterholes have been described and illustrated by DeVore and Washburn [1963], and Struhsaker [1967] has seen baboons and vervets drink at the same time from the same waterhole. Doubtless, all are safer as a result.

1. Impala

Relations between baboons and impala gazelle *(Aepyceros melampus)* are of particular interest because of the loose symbiosis between them. Baboons and impala were often seen foraging and progressing near each other—even intermixed with each other—with practically no interaction. In general, there seems to be a slight attraction between the two. In part, however, much of the association results from attraction of both species to a common part of the habitat: open, grassy areas, near or under fever trees. When either species becomes alarmed, the response spreads immediately to the associated animals of the other species. In this way, each species has the advantage of the particular capabilities of the other species to detect their common predators, such as leopard and lion. In addition, impala, unlike many of Amboseli's ungulates, do not migrate away from the vicinity of the permanent waterholes during the rainy season. Their association with baboons is thereby extended.

2. Grant's Gazelle

Unlike impala, Grant's gazelle *(Gazella gazella)* are frequently found in very open country (e.g. grassland with a few umbrella trees) and may

remain in the open in the hottest weather [LAMPREY, 1963], although in
Amboseli they sometimes take advantage of available shade. We have
only a few records of Grant's gazelle near or among baboons. In most
cases, they ignored each other. Once, a group of these gazelle that were
intermixed with the Main Group became alarmed and ran. The alarm
spread to the baboons, who ran in the same direction. On the other
hand, adult male baboons sometimes chased Grant's gazelle, and we
have described above a case of baboons preying upon a young gazelle
of this species.

3. Thomson's Gazelle

We have but one record of baboons in close proximity to Thomson's gazelle
(Gazella thomsonii). During a progression, the Main Group flowed around
two of these animals, neither of which seemed alarmed or concerned.

4. Warthogs

Relations between baboons and warthogs (Phacochoerus aethiopicus),
which the baboons sometimes encountered near waterholes, were hardly
cordial. Baboons encountered warthogs most frequently near permanent
waterholes. The two species tolerated each other at a distance of 30 ft.
or so, and their groups sometimes partially intermingled without altercation. At other times, warthogs chased baboons, even adult males.
When this happened, the baboons quickly got out of the way, but did
not bark or show other signs of alarm.

5. Vervet Monkeys

Of all the baboons' relations with other species, those with the vervet
monkeys are among the most fascinating. These two species exhibit a
combination of interrelations that may be unique in the animal king-
dom. First of all, baboons and vervets occasionally play with each other.
Large infant and small juvenile baboons, either singly or in small
groups, were observed playing with young vervets 3 or 4 times during
our study (fig.59). Body contact was seldom seen during this play;

Fig. 59. Young baboon and vervet in rough-and-tumble play.

rather, it usually consisted of running and reciprocal chasing, both on the ground and within the trees.

Second, each of these two species responds to the alarm calls of the other, and thus each assists the other to become aware of predators in the vicinity. We have observed simultaneous alarm-barking by baboons and vervets toward lion and leopard, and vervets, like baboons, are known to be preyed on by leopard [STRUHSAKER, 1967].

Third, there is niche overlap in the two species. They share food and water resources, and both species use groves of fever trees for sleeping [STRUHSAKER, 1967]. Indeed, on one occasion a group of vervet monkeys slept in Sleepy Hollow, the most frequently used sleeping grove of the Main Group, during the same night that the Main Group was in the grove. Niche overlap is doubtless considerably reduced by differences in habitat preference in the two species: unlike vervets, baboons readily move out onto open grassland, so far from trees that they cannot resort to them for safety. Vervets, in contrast, tend to trot when moving across large open expanses between trees.

Fourth, there are frequent agonistic encounters between baboons and vervets, despite the fact that vervets tend to avoid baboons, typically by ascending and remaining in trees when baboons approach. Baboons, from the size of a small juvenile upward, are able to chase off even the large adult male vervets. At times most members of a baboon group will bark at and chase vervet monkeys in a response that is very similar to their response to a predator or prey animal. Typically, the vervets responded to this by retreating to the heights of a tree. Finally, as we indicated in the chapter on baboon foods, baboons sometimes eat vervet monkeys.

Yet, despite these intensive interactions between the two species it is sometimes quite surprising to see a group of vervets move close to or even through a group of baboons and elicit nothing more than watchfulness and a few grunting vocalizations.

6. Other Mammals

Relations between baboons and the other large mammals of Amboseli, such as elephant, giraffe, rhinoceros, waterbuck, wildebeest, and zebra, are usually neutral, except that baboons move quietly out of the way if these larger animals move toward them. In addition, we have some observations that suggest that zebra and wildebeest may respond to baboon alarms, and vice versa.

7. Eagles and Buzzards

Once, a single baboon of the Main Group screeched as an African hawk eagle (*Hieraaetus spilogaster*) flew low over the sleeping trees at 0750. Yet one day, the baboons ignored an eagle of this species to which vervet monkeys were giving vocal threats, and on another day they only responded with cohesion grunts as vervets barked at a large bird, doubtless an eagle. When an unidentified eagle landed in a dead tree some 70 yards from the Main Group, cohesion grunts were given. And when an eagle (or vulture) flew directly into the sleeping trees, an adult male gave a deep, bass grunting vocalization. On another day, the baboons of the Main Group moved directly under a fever tree atop which sat two augur buzzards (*Buteo rufofuscus*) without showing any concern. When two vultures went through their noisy mating process one morning near a dozing adult male baboon, he ignored them.

8. Owls

Verreaux's eagle owl *(Bubo lacteus)* was ignored by the baboons. It is not known whether these or any other raptors of Amboseli ever prey upon baboons.

9. Starlings and Spurfowl

Baboons alerted when superb starlings *(Spreo superbus)* squawked at a slender mongoose, and when yellow-necked spurfowl *(Pternisis leucoscepus)* suddenly took flight, probably in response to a raptor.

10. Hornbills, Bustards

Baboons generally ignored ground hornbills *(Bucorvus leadbeateri)* and kori bustards *(Ardeotis kori)* when these large birds walked near or amongst them. Bustards, particularly, will unperturbedly walk through the midst of a baboon group. On one occasion, when a bustard was approached quite closely by baboons, it gave several, deep, grunt-like vocalizations, then walked on.

11. Rollers

At times, lilac-breasted rollers *(Coracias caudata)* remained near baboons and seemed somewhat attracted to them. If grasshoppers took flight when disturbed by the baboons, the rollers would sometimes catch them in mid-air. Young baboons that tried to catch a flying grasshopper would sometimes have it caught directly in front of them by a roller. Once, a large juvenile male baboon briefly chased a roller. On all other occasions, the baboons ignored these birds.

12. Plovers and Lapwings

Blacksmith plovers *(Hoplopterus armatus)* and crowned lapwings *(Stephanibyx coronatus)* often mobbed the baboons with screaming dives. Spotted eggs were found in open grassland on June 28; they were probably those of the crowned lapwing. A week later, the eggs were gone, and scrape marks through the dirt at that spot may have been left by the predator, which might have been a baboon. We have no direct evidence of predation by baboons on these birds or their eggs.

13. Comparisons and Discussion

Several authors comment on the fact that baboons associate, without altercation, with other animals—except those that they eat or that eat them.

'In nature it is comparatively rare to see baboons taking much notice of animals which do not normally serve as food. In the Kruger National Park it was a common feature that guinea fowls, impala, bushbok, blue duiker and baboons fed together without taking any notice of one another. On a few occasions, however, I have seen baboons tease bigger animals—on one a young female kudu, and on two others, lions. In the case of the kudu, a bush obstructed my view but, as far as I could see only young baboons participated. The game only consisted in lashing out with the hands at the kudu's legs, the result being that the poor animal which was completely surrounded, performed the most peculiar jumps on the spot, unable to escape either in one direction or another' [BOLWIG, 1959].

'An example of a presumably unusual antipathy between baboons and eland was seen at the Cape, 1961. A cow eland (one-horned), which was closely accompanied by two calves, on two occasions charged at S group, scattering the whole group. Although the dominant male on the second occasion then walked directly towards this cow and made threat intention-movements at her, there was no doubt of the group's general wariness of this particular animal, though not of others in the Reserve.... With the exception of the eland episodes referred to above, the relationship between baboon groups and all other ungulates has invariably been one of apparent mutual disregard. This applies to all the different species observed near baboons in S.W. Africa, S. Rhodesia, the Drakenbergs, and the Cape Reserve' [HALL, 1963].

Subsequently, HALL [1965a] noted bushbuck chasing baboons in Uganda. '...Baboons are more "casual" than patas, sometimes just stepping out of the way of feeding elephants when the latter are only a few metres from them' [HALL, 1965a].

'Antelope [in the Ishasha River valley, Uganda] include topi, kob, waterbuck and bushbuck but interaction between the baboons and these, and warthog and forest hog seemed to be minimal. Ungulates and baboons, though they took note of each others' alarm calls, rarely acted on them; there was none of the interspecific co-operation reported by DeVORE [1963]' [ROWELL, 1966].

'In the densely populated game reserves of Kenya [baboons] are seldom far from other animals.... The baboons' relationship with most of these species is mutual tolerance, and most of the time the species simply ignore each other' [DEVORE and HALL, 1965].

CROOK and ALDRICH-BLAKE [1968] and KUMMER [1968] have described a number of interactions between hamadryas baboons and gelada 'baboons' *(Theropithecus gelada)* in Ethiopia. Hamadryas were observed by CROOK and ALDRICH-BLAKE to walk directly through a group of geladas. In this, as in most of the other cases that are described, some of the geladas moved out of the way of the hamadryas, but otherwise took virtually no notice of them.

That baboons become alerted or alarmed in response to alarm responses of ungulates and vice versa has been noted in several of DEVORE's papers [e.g. WASHBURN and DEVORE, 1961; DEVORE and HALL, 1965; EIMERL and DEVORE, 1965]. HALL [1965a] noted such relations between patas monkeys and oribi in Uganda. STRUHSAKER [1967] records 7 species of ungulates in such relations with vervet monkeys, and comments that, 'It seems quite likely that such a relationship exists between vervets and all sympatric ungulates having similar predators, combining the advantages of the visual acuity of the monkeys and the olfactory sensitivity of the ungulates.'

The kinds of interrelations and interactions between baboons and vervets that we described above were confirmed by STRUHSAKER [1967] during his study of vervets in Amboseli:

'...Encounters between baboons and vervets were seen almost daily. In the great majority of these encounters there was spatial supplantation of vervets by the baboons. The influence of baboons on movements of vervets seemed greater in open than in wooded areas. The terminal ends of small branches were among the few places from which baboons could not supplant vervets, due to their large size. The relationship between these two species was extremely complex. Not only were they food competitors, but as mentioned earlier, baboons preyed on vervets. In addition, they sometimes slept together in the same tree, drank simultaneously from the same water hole, and warned one another of mutual predators such as leopard and lion. Furthermore, young baboons frequently approached vervets in a "playful" manner. The vervets responded by moving away and sometimes by jerking their forequarters toward the baboons in a threatening manner. After one such approach an agonistic encounter ensued in which a vervet and juvenile baboon actually grabbed at one another with their hands.'

1. The Mental Map

Direct movement of baboons over long distances to some of the larger, semi-permanent rain pools, even when the baboons could not see the rain pool during their progression, convinced us that they were familiar with the locations of these water sources. We believe that they were orienting on the basis of memory, rather than the immediate sensory perception of water. Similarly, they would come quite directly to a sleeping grove from different directions on different afternoons, even when intervening foliage probably made it impossible for them to see the grove from the ground.

We feel certain that the baboons were familiar with the major topography of their home range, and could navigate within it on the basis of that familarity. Such a system is obviously more efficient than random wandering, and is one of the factors that extends the dependence of the young baboon from a reliance solely upon his mother to a dependence upon the collective experience and action of the entire group.

In all likelihood, familiarity with a particular area results in greater efficiency in exploiting resources and in avoiding hazards. This would give a selective advantage to those behaviors—which might be called 'traditional'—that tend to bind an animal or group of animals to an abiding home range. This advantage should be most marked for animals with large home ranges, some very restricted resources, great longevity, and good memories—exactly the combination of traits found in savannah baboons.

2. Home Range Utilization

It is now clear that for a group of baboons, there is no single pattern of home range utilization. There is not one single progression route nor even a unique route between any two parts of their range, no single sleeping grove nor even a fixed sequence of rotation among sleeping groves, no one preferred food that is eaten to exhaustion before the next in order of preference is taken. There is no one waterhole that is used in preference to all others, and no invariable response to predators, even those of a single species. Regardless of what biologically important activity we examine, we find, not a fixed pattern of action, but rather, a distribution of responses.

In this connection, we can imagine two possibilities. Perhaps for each such class of activities there is some one response that is adaptive, all others being maladaptive. It might be supposed that the baboons, even though they are not perfect, utilize this adaptive response often enough to survive. Under such circumstances, selection would favor any population of baboons that made the appropriate response more frequently or more accurately. Alternatively, it may be that there are a variety of adaptive responses; if the adaptiveness of each course of action depends on the particular circumstances, the distribution of these circumstances will establish the distribution of adaptive responses.

Each portion of a group's home range can be regarded as having a certain utility to the animals, namely, the difference between what they gain from the area, in terms of access to those resources that will enhance survival, growth, and reproduction, and what they lose, as a result of hazards to life encountered therein.

In the short run, the utility of an area is not constant. It depends on variations in the environment and on previous behavior of the animals. But in the long run, each area will have a net utility. For example, the value of water sources to a group of baboons is very different, depending on whether it has been a cool day, or on whether the baboons recently drank[53]. But in the long run, baboons in a particular area have a net water loss that must be restored if they are to survive, and this, in turn, establishes a net utility of water resources for these animals.

The perpetuation of a group depends upon its ability so to allocate the distribution of its activities among the parts of the home range that the net utility of the range to the animals is positive, or at least, non-negative. Successful (i.e. well-adapted) social groups will be characterized by positive utilization rates. In such groups, birth rate will exceed death rate. In a stable population without intergroup migration, such

[53] Thus, the baboons' decision problems might be regarded as a stochastic game, in that 'the player's strategies control not only the...payoff but also the transition probabilities which govern the game to be played at the next stage' [LUCE and RAIFFA, 1957].

groups will inevitably expand more rapidly than others with lower (particularly, negative) utilization rates.

Thus, in analyzing the utilization of home range, we must consider not only those parts of the home range that the animals enter frequently or remain in for long periods of time, but also those that are seldom and briefly entered. More precisely, we must consider the spatial distribution of home range utilization and its relationships to the distribution of both hazards and natural resources among the parts of the home range.

In Amboseli, the following major natural resources of the baboons vary markedly from one part of the home range to another: (i) seasonal rain pools; (ii) permanent waterholes; (iii) groves of trees suitable for group sleeping; (iv) sources of plant foods; and (v) safe, short routes interconnecting the above.

Major hazards that vary from one part of the home range to another include: (i) intense insolation, lack of shade, and remoteness from water, with attendant dangers of dehydration and thermal imbalance; (ii) intergroup competition; (iii) mosquitos, schistosomiasis, and coxsackie B_2 (endemic to Amboseli baboons), all of which are probably most abundant or most readily transmitted in or near the permanent waterholes; and (iv) carnivores, which, for Amboseli baboons, include leopard, lion and silver-backed jackal, and probably also cheetah, spotted hyaena, striped hyaena, hunting dog, martial eagle and hawk eagle. Of these, the leopard is probably the baboons' major predator.

But if the utility to the baboons of a portion of their home range depends only on the balance between such risks on the one hand and the payoff to the animals on the other hand, then a particular degree of utility may result in some parts of the range from a high risk rate and a correspondingly high resource value, in others, from a low risk rate with a correspondingly low resource value, and similarly for intermediate values. For example, an area sparsely covered with edible grass and remote from any undergrowth that could conceal a leopard might have the same low but positive utility value to the baboons as another area in which free water is available and plant growth is lush but the chances of predator attack are high.

Beyond that, some hazards, such as attack by a leopard, materialize suddenly. Others, such as dehydration or thermal imbalance in areas without shade or water, develop relatively slowly. Similarly for essential resources: some, such as water, can be utilized rapidly by baboons, compared with others such as food, or, in the extreme case, sleeping trees. Among foods, fruits and large seed pods may be utilized relatively quickly; other foods such as the rhizomes of grass plants, which must be dug from the ground, can only be utilized more slowly. Thus, one must consider not only the frequency with which each part of the home range is entered, but the duration of occupancy, as well.

Areas with high risks or low payoff probably will be infrequently entered by well-adapted baboon groups. Similarly, unexpected hazards and quickly accessible resources should result in occupations of short duration, such as one gets, in Amboseli, in areas of high, dense *Sporobolus robustus* grass. Contrariwise, areas that are frequently entered and long-occupied probably have low, slow risks and high, prolonged payoffs; e.g. areas containing sleeping groves.

Such considerations have lead us to the concept of an adaptive distribution of baboon activities, which may be stated as follows: For any set of tolerable ecological conditions, the adaptive activities of baboons tend in the long run toward some optimal distribution away from which mortality rate is higher, or reproductive rate is lower, or both.

3. Home Range Overlap

The natural tendency of animals to occupy all available parts of the habitat while minimizing competition with conspecifics, combined with the advantages of an established, familiar area, tend to produce a mosaic distribution of home ranges, with contiguous or minimally overlapping boundaries. Such a situation cannot prevail, however, if one or more essential resources are not well-distributed throughout the habitat. When that is the case, we propose two general hypotheses to explain home range overlap. First, the amount of home range overlap will depend primarily upon those essential resources that have the most restricted spatial distribution. Consequently, home range overlap will be low in relatively uniform environments, and will be extensive where several essential resources have very restricted distributions. Second, the amount of time that groups with overlapping ranges will simultaneously be in the shared portion of their ranges will depend primarily on those essential resources in the overlap zone that can only be utilized slowly and whose availability is most restricted in time. As a result, simultaneous occupancy of overlap zones will be long wherever slowly utilized resources have a restricted period of availability, and conversely, will be brief if those resources that require longest to utilize are continuously available.

Thus, an essential nature resource is a restrictive factor in home range separation, in time or space, to the extent that increasing its dispersion in time or space will reduce home range overlap.

4. *Home Range Size*

For a mammal, body weight influences the rate of energy expenditure, and this, in turn, influences the amount of food that the animal must obtain to supply its requirements. McNab [1963] has pointed out that this results in a fairly linear correlation, in mammals, between body weight and home range size, with home range sizes usually being larger for 'hunters' (including those species that are either granivorous, frugivorous, insectivorous or carnivorous) than for 'croppers' (including those species that are either grazers or browsers) of the same weight.

McNab's correlation spanned mammals ranging in size from shrews to moose, and showed considerable 'scatter' about the least-squares regression line. Thus, in making intra-specific comparisons, within a relatively small size range, other factors may have an over-riding influence.

An animal's basic metabolic limitation is that energy output cannot indefinitely exceed energy input. While basal metabolic rate is closely related to body size, primarily as a result of the ratio of surface area to body mass, additional energy expenditure depends upon many factors that collectively might be called the animal's 'mode of life'.

We would expect available energy sources to be strictly proportional to home range size only if the habitat were uniform. But it never is, not only because of spatial variations in the food resources themselves, but also because of the effects of home range overlap. For a species in which population density is food-limited, the number or biomass of individuals that can be supported by an area depends upon the food productivity of the area, regardless of home range sizes and the extent of home range overlap. It seems plausible that these latter will be affected primarily by the distribution of resources and the strategies that are employed by the animals in exploiting them.

Ultimately, home ranges are limited to areas lying within cruising range of the essential resource with the most restricted distribution. Yet, for animals of many species, there must be large areas of land within the limits of their distribution in which even the most restricted resource is sufficiently well-dispersed that it places no such ultimate restriction on the animals' home ranges. If, despite this lack of restriction, home ranges do not drift, but tend to remain fixed in place, other factors, such as the advantages of a fixed abode, reinforced by competition from other home range units, must provide an explanation.

Whereas resources with restricted distributions place an upper limit, sparce resources place a lower limit on home range size[54]. The animals must cover enough ground to obtain a subsistence amount of those resources. Home range size is bounded by these two limits. Where these limits are close to each other, they will be the primary factors that determine home range size.

HALL [1963] suggested the possibility that in baboons the typical extent of day-journeys and the size of the home range were relatively constant and that baboon groups would grow 'to fit its viable limit'. HALL himself did not take this proposal too seriously, however, since he knew that it was at variance with a number of his own observations as well as those of others [HALL, ibid.]. The alternative that we suggest here is that home range size and the mean length of day-journeys depend primarily upon those vital resources with the sparsest and most restricted distributions. For example, in an area in which sleeping trees, water and other resources are abundant, but food plants are sparsely distributed, we would expect the lower limit on home range size to be determined primarily by available food.

This hypothesis could be checked observationally, by studying the correlation of home range size and of day-journey length with resource distribution, either in various habitats or in the same habitat with changing climate, e.g. seasonal or annual variations. Experimentally it could be checked by increasing the local abundance of the most sparsely distributed resource. If, for example, plant growth is limited by the availability of soil nutrients, these could be introduced locally [cf. WATSON and MOSS, 1971].

---

[54] Of course the sparseness of a resource depends not only on its density distribution, but also upon the needs of the animals.

5. *Population Dynamics and Group Size*

In chapter III, we reviewed the available data and ideas on the ecological correlates of baboon group sizes. Two quite different problems are
involved, first, to account for the distribution of group sizes within a
population, and second, to account for the fact that mean group size
varies greatly from one region to another. Put somewhat differently, we
need to account for both differences between populations and differences
within them. To date, the proposed explanations have all hinged on
gross correlations between group size and habitat type. Another, and
quite different approach involves studying the actual dynamics of
changes in group size, then relating these to underlying population
processes and to the effective ecological phenomena.

Regardless of the ecological correlates of group size, the distribution of group sizes in an area is a reflection of the dynamics of individual groups. The size of a group may change as the result of 6 processes:
birth, death, emigration, immigration, group splitting ('fission'), and
amalgamation of groups ('fusion'). The last of these processes has not
been reported in any free-ranging cercopithecine primate and is presumably so rare as to have no major effect on the distribution of group sizes.

The question may now be asked, what combination of birth, death,
emigration, immigration and group splitting results in an equilibrium
distribution of group sizes, and what is the nature of the distribution at
equilibrium?

A model that predicts the distribution of group sizes on the basis
of rates of birth, immigration, death and emigration—the BIDE model
—has recently been described by COHEN [1969], based on earlier work
by KENDALL [1949] and others. COHEN has applied this model, with
remarkable success, to the distribution of group sizes in several species
of primates.

The model is based on the following assumptions. During any
time·period, each individual in the population has certain probabilities
of giving birth to a single other individual, of dying, or of emigrating
from the social group, and these are the same for all individuals in the
group. Thus, the probability of a birth, a death, or an emigration in a
group during any time period will depend on group size. On the other
hand, immigration into the group is assumed to take place at a rate that
is independent of group size. These processes in the group are assumed
to be mutually independent, and each group is assumed to develop

independently of every other group. Under these assumptions, the probability $p(k)$ that the population will contain k individuals after a long time period is independent of starting size and is given by the negative binomial distribution

$$p(k) = \binom{r + k - 1}{k} p^r q^k, \quad k = 0,1,2\ldots,$$

where r is the ratio of the emigration rate to the birth rate, $p = 1 - q$, and q is the ratio of the birth rate to the 'loss rate' (death rate plus emigration rate).

The fit of this model to data on group size distributions in chacma, anubis and yellow baboons is shown in table XXV, which is reproduced with permission from COHEN's manuscript[55]. The fit to HALL's chacma data is poor, perhaps because data from several areas were pooled. The fit to our data from Amboseli is good, as is the fit when our data are combined with WASHBURN's and WARSHALL's from Amboseli.

In fitting these models to data on group size distributions to obtain the expected values shown in table XXV, the ratio of immigration rate to birth rate, and of birth rate to loss rate were estimated from the size distribution. Independent evaluation of these parameters of the model were made by COHEN from data on one group of baboons, our main study group in Amboseli. Using the data in table IV herein, COHEN estimated the birth rate at 6.49×10^{-4} births per individual per day, the immigration rate as 8.043×10^{-3} immigrants per group per day, and the loss rate at 9.74×10^{-4} deaths or emigrations per individual per day. The expected distribution (table XXVI herein, reproduced with permission from COHEN, 1969, table IX) departs markedly from our data and from a combination of our data and those of WASHBURN and WARSHALL from Amboseli.

The poor fit of the model using these estimates of the parameters may be attributable to any of several sources [cf. COHEN, 1969]. Estimates of the birth-immigration-death-emigration parameters from this one group may be biased as a result of sampling error. Or the parameters of this group may be atypical. These are empirical problems for which the solution depends on getting more field data than are now available.

[55] COHEN [ibid.] also fits this model, and a related one based on the Poisson distribution, to published data on group sizes in colobus monkeys, langurs, howler monkeys, and gibbons.

Table XXV. Frequency distributions of sizes of baboon troops, observed (from literature and this publication) and predicted by the BIDE model (from COHEN [1969])

I Size	II HALL	III Pred.	IV ALT-MANN	V Pred.	VI WASH-BURN	VII WAR-SHALL	VIII Yellow	IX Pred.	X DE VORE	XI WAR-SHALL	XII Olive	XIII Pred.	XIV Total	XV Pred.
1– 10	1	3.9	…	1.7	…	1	1	3.2	…	…	…	3.1	2	10.2
10– 20	12	10.2	2	5.3	1	2	5	8.1	2	…	2	3.5	19	21.0
20– 30	16	11.3	11	7.5	…	3	14	10.6	3	…	6	3.5	36	24.1
30– 40	9	9.4	10	7.9	…	2	12	11.1	…	…	3	2.8	24	22.6
40– 50	4	6.8	7	7.2	2	3	12	10.3	1	…	2	2.0	18	19.1
50– 60	6	4.5	7	5.9	2	…	10	8.8	…	…	…	1.3	16	15.2
60– 70	1	2.8	4	4.6	2	1	6	7.3	1	…	2	…	9	11.6
70– 80	1	1.7	3	3.4	3	1	7	5.8	1	…	2	…	10	8.6
80– 90	1	…	5	2.4	1	1	7	4.5	1	…	2	1.8	9	6.3
90– 100	2	2.3	…	1.7	1	1	2	3.4	…	…	1	…	4	4.5
100–110	…	…	1	1.2	1	1	2	2.5	…	…	…	…	…	3.1
110–120	…	…	…	…	…	…	…	1.8	…	…	…	…	…	2.2
120–130	…	…	…	…	…	…	…	1.3	…	…	…	…	…	1.5
130–140	…	…	…	…	…	…	…	…	…	…	…	…	…	1.0
140–150	…	…	…	…	…	…	1	1.6	…	…	…	…	1	…
150–160	…	…	…	2.2	…	1	1	…	…	…	…	…	1	…
160–170	…	…	…	…	1	…	1	…	…	…	…	…	1	…
170–180	…	…	…	…	1	…	…	1.6	…	…	…	…	1	2.1
180–190	…	…	…	…	…	…	…	…	…	…	…	…	…	…
190–	…	…	1	…	…	…	1	…	…	…	…	…	1	…

	III Pred.	V Pred.	IX Pred.	XIII Pred.	XV Pred.
$\hat{p}$	0.0727	0.0560	0.0435	0.0829	0.0470
$\hat{r}$	2.729	2.986	2.518	3.717	2.296
χ^2	7.827	11.381	15.305	5.495	22.961
df	6	9	12	4	12
P	(0.2, 0.3)	(0.2, 0.3)	(0.2, 0.3)	(0.2, 0.3)	(0.02, 0.05)

Data are from DEVORE and HALL [1965], this publication, and WARSHALL [unpublished], Unpublished data reproduced by kind permission of Mr. PETER J. WARSHALL. *I* Size of baboon troops; for choice of intervals, see COHEN, 1969. *II* HALL's observations of chacma baboons, southern Africa. *III* Truncated negative binomial distribution fitted to II. *IV* Our observations of yellow baboons in Amboseli, Kenya. *V* Truncated negative binomial distribution fitted to IV. *VI* WASHBURN's observations of yellow baboons in Amboseli, Kenya. *VII* WARSHALL's observations of yellow baboons in Amboseli, Kenya (July 1964). *VIII* Sum of IV, VI, and VII (all yellow baboons). *IX* Truncated negative binomial distribution fitted to VIII. *X* DEVORE's observations of olive baboons, Nairobi Park, Kenya. *XI* WARSHALL's observations of olive baboons, Nairobi Park, Kenya (1964). *XII* Sum of X and XI (all olive baboons). *XIII* Truncated negative binomial distribution fitted to XII. *XIV* Sum of II, VIII, XII (all baboons). *XV* Truncated negative binomial distribution fitted to XIV. With 14 df. $0.05 < P < 0.10$.

Table XXVI. Observed and predicted distribution of group sizes in Amboseli when the parameters ($p = 0.333, r = 12.3$) of the BIDE model are estimated from data in table IV

Size	I ALTMANN only observed	predicted	II All Amboseli observed	predicted
1–10	..	1.4	1	2.2
10–20	2	16.0	5	25.7
20–30	11	21.8	14	35.1
30–40	10	9.4	12	15.2
40	28	2.4	50	3.8
χ^2	297.15		592.46	
df	2 or 4		2 or 4	
P	<<0.01		<<0.01	

The model is fitted to our Amboseli data (fig. 6), and to a combination of our data with those of WASHBURN and of WARSHALL (table XXV) for Amboseli baboons.

Another set of problems stems from the model itself: it may not be an adequate representation of baboon group dynamics. A statistical fit alone is not sufficient grounds for accepting a model: if some assumptions of the model are known to be unrealistic, we must search for other models that fit the data as well, but that are based on more realistic premises. (This is particularly important when distributions such as the negative binomial are used. This distribution is known to fit unimodal, asymmetric data from a wide variety of sources.)

The BIDE model might be elaborated by taking some of the following factors into account.

(i) Birth and death processes are probably not independent, either of each other, or from individual to individual. For example, death of the mother of a dependent infant reduces the infant's chances of surviving

(ii) Birth rate must be more closely related to the number of individuals of reproductive age in the group than to total group size. In particular, the numbers of adult females in the group is probably a much better census datum to use than group size when estimating the reproductive capacity of the group. Estimates of reproductive rate per female have been given on page 59.

(iii) Mortality rate is surely not the same for all classes of animals. To judge by data from other vertebrates, the age-specific mortality rate (i. e. the probability that an animal of age x will not live to $x + 1$) is high during infancy, then relatively low and level thereafter until the late adult period, during which it is again high. Our estimates of death rates are given on page 62.

(iv) Our data on migrants suggest that intergroup migration is very unlikely for any class other than fully adult males. Thus, the number of adult males per group, rather than total group size, should be used in estimating the probability of an emigration during any time period. Estimates of emigration rate per adult male have been given on page 63. As for immigrants, our data (chapt. III) do not contradict the assumption that the probability of immigration is independent of group size.

(v) Group fission has not been taken into account, yet it has been reported several times in the baboon-macaque group [FURUYA, 1960; KOFORD, 1966; SUGIYAMA, 1960, and chapt. III herein].

(vi) The model treats each group as an independent replication, unaffected by the outcome of other groups. Yet, this is surely not the case. For example, one group's immigrant must have been another group's emigrant. In a closed system of groups, the number of immigrants before any point in time cannot exceed the number of emigrants before that time, and if, as in the BIDE model, emigration rate depends upon group size but immigration rate does not, then small groups will grow at the expense of large ones, and thus intergroup migration will tend to equalize group size. But with a model that assumes independence of groups, no group expands at another's expense; the effect of immigration and emigration on group size depends solely upon the balance between the two, and that balance is unaffected by what happens elsewhere.

(vii) The model is an equilibrium model that assumes constant, time-independent rates of birth, death, and migration. We find it difficult to believe that these rates have not varied widely, even within quite recent times, as a result of major fluctuations in climate. Severe drought gripped East Africa from late 1959 until late 1961. The drought was followed by exceptionally heavy rains that flooded many areas. Such rains occurred again during 1968. The vegetation is dramatically altered by such climatic changes [BUETTNER-JANUSCH, 1965]. It seems likely that such changes in the environment will have marked effects on baboon populations.

From what we have said above, it should be clear that improving the BIDE model is going to require not only further mathematical developments, but also, more and better data on primate populations than are now available.

6. Sexual Dimorphism and the Sex Ratio

DeVore and Washburn [1963] have speculated on the ecological significance of the difference in size between adult male and adult female baboons:

'The role of the adult male baboons as defenders of the troop has been described. This behavior is vital to the survival of the troop, and especially to the survival of the most helpless animals—females with new babies, small juveniles, and temporarily sick or injured individuals. Selection has favored the evolution o males which weigh more than twice as much as females, and the advantage to the troop of these large animals is clear, but it is not obvious why it is advantageous for the females to be small. The answer to the degree of sex differences appears to be that this is the optimum distribution of the biomass of the species. If the average adult male weighs approximately 75 pounds and the average adult female 30 pounds, each adult male requires more than twice the food of a female. If the food supply is a major factor in limiting the number of baboons, and if survival is more likely if there are many individuals, and if the roles of male and female are different—then selection will favor a sex difference in average body size which allows the largest number of animals compatible with the different social roles in the troop.

If selection favors males averaging 75 pounds, then it will favor females which are as much smaller as is compatible with their social roles. Since the females must travel the same distances, carry young, engage in sexual and competitive activities, there are limits to the degree of sexual differentiation, but the adaptive value of the difference is clear. For example, a troop of 36 baboons composed of 6 adult males and 12 adult females and their young (18 juveniles and infants) has a biomass of some 1,000 pounds. If the females also weighed 75 pounds each, 6 adult males and 6 adult females would alone total 900 pounds and have only one-half the reproductive potential of 6 adult males and 12 adult females. Because this would halve the number of young, it would greatly reduce the troop's chances of survival[56]. Our data are not sufficiently detailed to analyze the actual distribution of biomass in the troops we observed, but our observations are compatible with the limited data on weights and the numbers of adult animals we saw. Viewing sexual differentiation in size as a function of the optimum distribution of biomass of the troop offers a way of understanding sexual dimorphism fundamentally different from the view which considers only sexual selection, dominance, and intratroop factors. Obviously, all factors should be considered[57].

[56] DeVore and Washburn here seem to be postulating the action of group selection [cf. Wynne-Edwards, 1962], a process that presupposes special conditions that are unlikely to be found in nature [Maynard Smith, 1964]. All that is required here is that small females have a higher reproductive rate than large females. Evolution of the male's altruistic but hazardous behavior of actively defending the group against attack may be due, in part, to kin selection [Hamilton, 1963, 1964].

[57] While such selection may account for the evolution of dimorphism in size, it cannot account for other components of sexual dimorphism in baboons.

Adaptation is a complex process and results in compromises between the different
selective pressures, but a distribution of biomass which doubles the reproductive
potential of a species is so important that other factors may be minimized.'

This ecological argument may also be applied to the number of adult
males, and there may be ecological advantage in having females small
but numerous, and males large but relatively less numerous—yet still
sufficiently abundant that they can defend the group and fertilize the
females. If so, then such selective advantages may, in part, account for
the preponderance of females in most baboon groups and for a mating
system in which a male may breed with more than one female. KUMMER
[1968, p.21] noted a steady increase in the percent of adult males in
hamadryas troops as he moved from west to east: there were 15.05%
adult males in hamadryas troops in Aliltù and 22.6% in those near
Harar, Ethiopia. Correspondingly, the percent of adult females decreased from 36.5 to 30.8%. He offered no explanation for the gradient.
Food was more abundant and group size was larger near Harar [ibid.,
p.19]; correspondingly, there may have been less intense selection for
a disparate sex ratio in that area than in Aliltù.

7. Baboons, Vervets and Patas Monkeys

In a broad belt of savannah running all the way across Africa, primarily
north of the Equator, baboons encounter vervet monkeys and patas
monkeys [TAPPEN, 1960; HALL, 1965a]. All three occur in the Amboseli
Reserve. The 3 species are separated by major differences in their
niches, which, in simplified form, are these. Vervets are primarily
animals of open forest. They seldom and only briefly move far from
groves of trees. Patas are animals of open bush and grassland; they
seldom enter even those gallery forests that cut through their habitat.
Baboons get the best of both worlds: they make use of the forests for
shade, food, water, and sleeping trees, yet they are able to tap the immense food resources of the open grasslands at such great distances
from trees that they could not rely upon them for safety.

The late K.R.L. HALL had the unique combination of extensive
field experience with all three species in several areas of Africa. His
monograph on patas monkeys [HALL, 1965a] includes a number of
interesting comparisons between these three monkeys:

'The difference in the home range pattern as between baboons, vervets, and patas was made specially clear from seeing groups of the three species in the Chobi-Karuma area.... The baboon and the vervet group returned regularly to the same clusters of trees along the Nile bank every night on which observations were made—and were still doing so when GARTLAN [personal communication] spent 3 weeks in the area in July 1964, a year later. The patas group passed each night in a different area of the woodland savannah, and never went nearer to the bank than about 400 m in the course of its day-ranges. No patas group has so far been encountered in Murchison near the riverine vegetation or in the heavily wooded parts of some of the valleys. The baboon groups near Paraa likewise regularly returned either to tree clusters along the Nile or, in one case, to a tree cluster by the Emi river. Baboon groups range far into the savannah from these home bases, but there are many areas of savannah and erosion valley ranged by the patas where baboons were never seen. Thus, in the 28 consecutive days of observation of Group IV, not a single baboon group was sighted, only one isolate adult male baboon being seen twice in the home range. Patas Group II, on the other hand, met baboon groups several times in the southern third of its home range, but never in the northern two-thirds which was further away from the Nile. Vervets were often seen in the Chobi woodland savannah, but the pattern the one group showed was similar, on a much smaller scale, to that of the baboons, namely ranging out during the day into the savannah, then returning to the river bank. This particular group never went far from trees, and rushed back to trees whenever it was frightened. It tended to feed and rest during the latter part of the day along the Nile bank. Small parties of vervets were sometimes found far into the grass savannah away from the river, but they were always near small gullies with a few trees. Thus, although baboons and vervets overlap with the patas home ranges, and eat some of the same foods such as berries and fruits, there are large areas of *Combretum* bush grass savannah and erosion valley where only patas have been seen. The vervet still remains a tree or thicket-based monkey, in spite of its sometimes quite long grassland excursions. The patas, one may say, is essentially, by contrast, a ground-based monkey who sometimes uses trees or thickets for food and shade. The escape behaviour of the two species usually shows this clearly—patas running down trees and away across the ground, even in woodland, vervets running up into trees' [pp. 37–38].

'On the rare occasions when a baboon group approached directly an area where patas were feeding, the latter withdrew rapidly.... No threat behaviour or vocalizations have occurred with reference to the other, either on the side of the patas or of the baboons on these occasions. No doubt the patas know the probable behaviour of the baboons encountered in their home range with some accuracy, so that they do not unnecessarily take evasive action.

The only definite encounters with vervet groups occurred in the Woodland savannah at Chobi and Pamdera. There is no evidence that vervets withdraw from patas, or vice versa.... [Once a patas group] and a vervet group were between 80 and 150 m from each other for nearly an hour. During this period, patas and vervet watched each other, but there were no vocalizations and no alarm or threat behaviour. The vervets began to feed on the figs in one of the trees, these being one of the main food items at this time of the patas group. In the

more open grassland habitat, where vervets were rarely seen, no encounters with
patas were observed. On two occasions, volleys of calls that were probably from
vervet groups were heard, apparently when the patas group that we were following had come into contact with them. On both occasions, the adult male patas
barked repeatedly—a rare occurrence that otherwise was recorded only when a
patas group encountered other patas. As the behaviour of the vervets was not
observed, the data are inconclusive, and there remains from the present study
insufficient evidence as [to] the usual relationship between the two species.

Because of their way of life which, at Murchison and in West Africa [BOOTH,
1956], brings them only to the edges of forest, and of riverine thicket and tree
clumps, patas are very unlikely to encounter monkey species other than baboons
and vervets' [pp. 48–49].

'The fact that these patas groups went so rarely to the water sources in the
valleys, even when they were ranging close to them, suggests that there may be
some special hazard for them associated with such places. This is also indicated
by the watchfulness of the adult male, and by the briefness of the drinking.
Baboons showed no such caution at Murchison, some of them being seen to drink
from pools or wallows on all the full day-ranges in the Paraa area' [p.53].

'The patas has been supposed to be much less dependent on water availability
than baboons. BIGOURDAN and PRUNIER [1937] report the occurrence of patas
during the dry season in areas of West Africa completely devoid of water, whereas
the baboons at that season remain fairly close to water because they are said to
need to drink everyday.

At Murchison, apart from observations of a special type of drinking seen in
Group I in Pamdero...during the whole of the Stages I and II study period a
group, or individuals from a group, was seen to go to water holes, wallows, pools
or puddles to drink on only four occasions...' [p.52].

'Our present evidence certainly strongly suggests very clear and important
differences in the ecology and behaviour of patas and vervet, as well as in that
of patas and baboon.

We have said that the patas has evolved to extreme form amongst monkeys a
particular complex of adaptations to ground living. This complex contrasts in
many important features from that which makes up the kind of ground plan of
baboons and macaques. The major contrast seems to lie in the physical adaptation
for high speed ground locomotion, and the associated adaptations of evasion by
concealment, by dispersion, and by silence. Correlated with these is the uniquely
interesting watchful and perhaps diversionary behaviour of the adult male, and
the probably closely-organized social system of the adult females who, with the
young animals of the group, never take part in such behaviour but remain quite
separate from the adult male during these activities. The very small amount of
aggression, and, when it does occur, its usually non-contact variety, compares
very markedly with the baboon group the adult males of which can be exceedingly
rough in "disciplining" the group and exceedingly effective in combined defence-
attack against predators (see WASHBURN and DEVORE, 1960, film entitled
Baboon Behavior)' [p.83].

'The importance of the contrast between the patas and baboons is in emphasizing
the quite different survival pattern of the two genera. Baboons, being based on

rocky escarpments or cliffs or on large tree clumps, use the savannah or woodland from these bases, and rely for daytime security on the defensive power of adult males and on the alertness of peripheral males. Patas are truly creatures of the savannah itself, and there are no areas within the home ranges which are specially favoured because of the security they offer' [p.84]. HALL's paper includes a detailed tabular comparison of these two species [ibid., table XI].

An ecological comparison of baboons and vervets has been made by STRUHSAKER [1967]. STRUHSAKER's field work on vervets was carried out in the same area and at the same time as our baboon study. He wrote as follows:

'One of the more important aspects of the complex relationship between Amboseli baboons and vervets is that of niche separation, especially as it relates to food. Although baboons were physically capable of eating all the foods that vervets did, the converse was not true. Apparently, the smaller vervets lacked the strength or ability to pull and dig grass plants from the ground and thus expose their root systems. During the dry season, which prevails for ½ of each year the major food in the open-plains habitat for non-ruminants are grass roots. It is this food source which baboons so efficiently exploited. Thus, although these two cercopithecines did complete with one another for food, such competition was greatly reduced when baboons spent large amounts of time in the open-plains habitat utilizing a food that was unavailable to vervets.

Correlated with their relatively small size vervets demonstrated extreme arboreal agility, which was their major defense against diurnal predators. Furthermore, because of their small size, vervets had about four times as many predators as baboons. It may be this defense mechanism, linked with their relatively small size, inability to procure grass roots, and large number of predators that restricts vervets to the proximity of trees and thickets.

Related to this partial niche separation was the radically different predator defense system of baboons. Their defense depended on large body size and large canines, especially evident in the adult males, and in their large social groups. In Amboseli the baboons formed significantly larger (0.001 level) groups than did vervets (STRUHSAKER, in prep.). Such a defense permitted utilization of open habitat away from trees and thickets and, thus, further enhanced niche separation from the vervets.'

Although baboons are found in environments ranging all the way from moist, evergreen forest to semi-desert steppe, most of them live in savannah habitat, that is, in areas in which perennial grasses are the dominant ground cover, and in which trees, though of variable density, are never completely absent. Such plant associations cover much of the African continent [KEAY, 1959]. Thus, much of the abundance and wide distribution of baboons is the result of being successful in a predominant habitat. Nor is the present wide distribution of baboons something new: *Parapapio jonesi* of the early Pleistocene ranged at least from northwest Kenya to South Africa [PATTERSON, 1968].

The survival of baboons in open savannahs, often too far from trees to rely on them for safety, depends on a combination of behavioral, social and anatomical traits that enables them to defend themselves from attack by terrestrial predators [DEVORE and WASHBURN, 1963]. This aggressive potential makes it possible for them to displace any other monkeys that they encounter.

On the savannahs, the success of baboons depends upon their ability to exploit a wide variety of plant and animal food sources, and to feed selectively on some of the most concentrated sources of nutrients in their environment. The efficiency of this exploitation and indeed the survival of the animals depends, in turn, on the fact that through intimate familiarity with one particular area, the baboons of a group are able so to distribute their activities that they have adequate access to the essential natural resources of their home range without exposing themselves to excessive risks.

REFERENCES

ALTMANN, S. A.: Mathematical models of random subgroup formation. Abstract. Amer. Zool. 5: 712 (1965); The structure of primate social communication; in ALTMANN, Social communication among primates, pp. 325–362 (University of Chicago, Chicago/London 1967); Sociobiology of rhesus monkeys, III: The basic communication network. Behav. 32: 17–32 (1968).

ALTUM, J. B. T.: Der Vogel und sein Leben; 7th ed. (Niemann, Münster 1903).

ANSELL, W. F. H.: Mammals of Northern Rhodesia (The Government Printer, Lusaka, Northern Rhodesia 1960).

BARTLETT, D. and BARTLETT, J.: Observations while filming African game. Sth afr. J. Sci. 57: 313–321 (1961).

BERT, J.; AYATS, H.; MARTINO, A. et COLLOMB, H.: Le sommeil nocturne chez le babouin Papio papio. Observation en milieu naturel et données électrophysiologiques. Folia primat. 6: 28–43 (1967a); Note sur l'organisation de la vigilance sociale chez le babouin Papio papio dans l'est senegalais. Folia primat. 6: 44–47 (1967b).

BIGOURDAN, J. et PRUNIER, R.: Les mammifères sauvages de l'Ouest Africain et leur milieu (Lechevalier, Paris 1937).

BOLWIG, N.: A study of the behaviour of the chacma baboon, Papio ursinus. Behav. 14: 136–163 (1959).

BOOTH, A. H.: The distribution of primates in the Gold Coast. J. w. afr. Sci. Ass. 2: 122–133 (1956).

BRAIN, C. K.: Observations on the behaviour of Cercopithecus monkeys. Ann. N.Y. Acad. Sci. 102: 477–487 (1965).

BROWN, L.: Africa. A Natural History (Random House, New York 1965).

BROWN, L. E.: Home range and movement of small mammals. Symp. zool. Soc. Lond. 18: 111–142 (1966).

BUETTNER-JANUSCH, J.: Biochemical genetics of baboons in relation to population structure; in VAGTBORG, The baboon in medical research, vol. 1, pp. 95–110 (University of Texas, Austin 1965); A problem in evolutionary systematics: nomenclature and classification of baboons, genus Papio. Folia primat. 4: 288–308 (1966).

BURT, W. H.: Territoriality and home range concepts as applied to mammals. J. Mammal. 24: 346–352 (1943); Territoriality. J. Mammal. 29: 207–225 (1949).

BUXTON, A. P.: Further observations of the nightresting habits of monkeys in a small area on the edge of the Semliki Forest, Uganda. J. Anim. Ecol. 20: 31–32 (1951).

CALHOUN, J. B.: The social use of space. Physiol. Mammal. 1: 1–187 (1964).

CALHOUN, J. B. and CASBY, J. U.: Calculation of home range and density of small mammals. US Publ. Hlth Monogr. 55 (1958).

CARPENTER, C. R.: A field study of the behavior and social relations of howling monkeys (Alouatta palliata). Comp. Psychol. Monogr. 10: 168 p. (1934); A field study in

Siam of the behavior and social relations of the gibbon *(Hylobates lar)* Comp. Psychol. Monogr. *16:* 212 p. (1940).

CARTMILL, M. and TUTTLE, R. H.: Mammalian social patterns in a savannah environment. Abstract. Amer. J. phys. Anthrop. *25:* 202 (1966).

CITTERS, R. L. VAN; SMITH, O. A.; FRANKLIN, D. L.; KEMPER, W. S. and WATSON, N. W.: Radio telemetry of blood flow and blood pressure in feral baboons: a preliminary report; in VAGTBORG, The baboon in medical research, vol. 2, pp. 473–492 (University of Texas, Austin/London 1967).

COHEN, J. E.: On estimating the equilibrium and transition probabilities of a finite-state Markov chain from the same data. Biometrics *24:* 185–187 (1968); Natural primate troops and a stochastic population model. Amer. Naturalist *103:* 455–477 (1969).

COX, D. R. and LEWIS, P. A. W.: The statistical analysis of series of events (Wiley, New York 1966).

CROOK, J. H. and ALDRICH-BLAKE, P.: Ecological and behavioral contrasts between sympatric ground dwelling primates in Ethiopia. Folia primat. *8:* 192–227 (1968).

DALKE, P. D.: The cottontail rabbits of Connecticut. St. Conn. geol. nat. Hist. Surv. Bull. *65:* 1–97 (1942).

DALKE, P. D. and SIME, P. R.: Home and seasonal ranges of the eastern cottontail in Connecticut. Trans. n. amer. Wildl. Conf. *3:* 659–669 (1938).

DART, R. A.: The predatory transition from ape to man. Int. Anthrop. linguist. Rev. *1:* 201–217 (1953); The osteodontokeratic culture of *Australopithecus prometheus.* Transvaal Mus. Mem. *10* (1957); The carnivorous propensity of baboons. Symp. zool. Soc. Lond. *10:* 49–56 (1963).

DASMANN, R. F. and TABER, R. D.: Behavior of Columbian black-tailed deer with reference to population ecology. J. Mammal. *37:* 143–164 (1956).

DEVORE, I.: The social behavior and organization of baboon troops; anthrop. diss. Chicago (1962); Mother-infant relations in free-ranging baboons; in RHEINGOLD, Maternal behavior in mammals, pp. 305–335 (Wiley, New York 1963); Changes in population structure of Nairobi Park baboons, 1959–1963; in VAGTBORG, The baboon in medical research, vol. 1, pp. 17–28 (University of Texas, Austin 1965).

DEVORE, I. and HALL, K. R. L.: Baboon ecology; in DEVORE, Primate behavior: Field studies of monkeys and apes, pp. 20–52 (Holt, Rinehart and Winston, New York 1965).

DEVORE, I. and WASHBURN, S. L.: Baboon behavior. 16 mm sound, color film (University of California, Berkeley 1960); Baboon ecology and human evolution; in HOWELL and BOURLIÈRE, African ecology and human evolution. Viking Fund Publ. Anthrop., Chicago *36:* 335–367 (1963).

DICE, L. R. and CLARK, P. J.: The statistical concept of home range as applied to the recapture radius of the deermouse *(Peromyscus).* Contrib. Lab. vert. Biol., Univ. of Michigan *62:* 15 p. (1953).

DOUGALL, H. W.: DRYSDALE, V. M. and GLOVER, P. E.: The chemical composition of Kenya browse and pasture herbage. E. afr. Wildl. J. *2:* 82–125 (1964).

DOUGALL, H. W. and GLOVER, P. E.: On the chemical composition of *Themeda triandra* and *Cynodon dactylon.* E. afr. Wildl. J. *2:* 67–70 (1964).

DRAKE-BROCKMAN, R. E.: The mammals of Somaliland (Hurst & Blackett, London 1910).

EIMERL, S. and DEVORE, I.: The primates (Time, New York 1965).

FISHER, R. A.: The genetical theory of natural selection (Clarendon Press, Oxford 1930).

FITZPATRICK, SIR J. P.: Jock of the Bushveld (Longmans Green, London 1907).

FITZSIMONS, F. W.: The natural history of South Africa, vol. 1 (Longmans Green, London 1919).

FREUND, J. E.: Modern elementary statistics (Prentice-Hall, New York 1952).

FURUYA, Y.: An example of fission of a natural troop of Japanese monkeys at Gagyusan. Primates *2:* 149–179 (1960).

GILBERT, C. and GILLMAN, J.: Pregnancy in the baboon *(Papio ursinus).* Sth afr. J. med. Sci. *16:* 115–124 (1951).

GILLMAN, J. and GILBERT, C.: The reproductive cycle of the chacma baboon *(Papio ursinus)* with special reference to the problems of menstrual irregularities as assessed by the behavior of the sex skin. Sth afr. J. med. Sci. *11:* 1–54 (1946).

GOODALL, J.: Feeding behaviour of wild chimpanzees. Symp. zool. Soc. Lond. *10:* 39–47 (1963).

GOODALL, J. van LAWICK: New discoveries among African chimpanzees. Natl. Geogr. *128:* 802–813 A (1965).

GRAFTON, R. N.: Food of the black-backed jackal. A preliminary report. Zool. Afr. *1:* 41–53 (1965).

GRIFFITHS, J. F.: The climate of East Africa; in RUSSELL, The natural resources of East Africa (Hawkins, Nairobi 1962).

GRZIMEK, M. and GRZIMEK, B.: A study of the game of the Serengeti Plains. Z. Säugetierk. *25–61* (1960).

HADDOW, A. J.: Field and laboratory studies on an African monkey, *Cercopithecus ascanius schmidti* Matschie. Proc. zool. Soc. Lond. *122:* 297–394 (1952).

HALL, K. R. L.: Social vigilance behavior of the chacma baboon, *Papio ursinus.* Behaviour *16:* 261–294 (1960); Numerical data, maintenance activities, and locomotion in the wild chacma baboon, *Papio ursinus.* Proc. zool. Soc. Lond. *139:* 181–220 (1962a); The sexual, agonistic and derived social behaviour patterns of the wild chacma baboon, *Papio ursinus.* Proc. zool. Soc. Lond. *139:* 283–327 (1962b); Variations in the ecology of the Chacma baboon. Symp. zool. Soc. Lond. *10:* 1–28 (1963); Behaviour and ecology of the wild Patas monkey, *Erythrocebus patas,* in Uganda. J. Zool. *148:* 15–87 (1965a); Ecology and behaviour of baboons, patas, and vervet monkeys in Uganda; in VAGTBORG, The baboon in medical research, vol. 1, pp. 43–61 (University of Texas, Austin 1965b); Distribution and adaptations of baboons. Symp. zool. Soc. Lond. *17:* 49–73 (1966).

HALL, K. R. L. and DEVORE, I.: Baboon social behavior; in DEVORE, Primate behavior, pp. 53–110 (Holt, Rinehart and Winston, New York 1965).

HAMILTON, W. D.: The evolution of altruistic behavior. Amer. Nat. *97:* 354–356 (1963); The genetical evolution of social behaviour, I and II. J. theoret. Biol. *7:* 1–52 (1964).

HAYNE, D. W.: Calculation of size of home range. J. Mammal. *30:* 1–18 (1949).

HILL, W. C. O.: Taxonomy of the baboon; in VAGTBORG, The baboon in medical research, vol. 2, pp. 3–12 (University of Texas, Austin/London 1967).

HOPWOOD, A. T.: The generic names of the mandrill and baboons, with notes on some of the genera of Brisson (1762). Proc. zool. Soc. Lond. *117:* 533–536 (1947).

HOWARD, H. E.: The British warblers, a history, with problems of their lives, 6 vol. (Cambridge University, Cambridge 1907–1914); Territory in bird life (Murray,

London 1920); An introduction to the study of bird behavior (Cambridge University, Cambridge 1929).

ISEMONGER, R.M.: Snakes of Africa, Southern, Central and East (Thomas Nelson, Johannesburg 1962).

ITANI, J.: Postscript by the editor. Primates 8: 295–296 (1967).

JAY, P.: The common langur of North India; in DEVORE, Primate behavior. Field studies of monkeys and apes, pp. 197–249 (Holt, Rinehart and Winston, New York 1965).

JEWELL, P.A.: The concept of home range in mammals. Symp. zool. Soc. Lond. 18: 85–109 (1966).

JOLLY, C.A.: Introduction to the Cercopithecoidea, with notes on their uses as laboratory animals. Symp. zool. Soc. Lond. 17: 427–457 (1966); The evolution of baboons; in VAGTBORG, The baboon in medical research, vol. 2, pp. 23–50 (University of Texas, Austin/London 1967).

JORGENSEN, C.D.: Home range as a measure of probable interactions among populations of small mammals. J. Mammal. 49: 104–112 (1968).

KALTER, S.S.; RATNER, J.J.; RODRIQUEZ, A.R. and KALTER, G.V.: Microbiological parameters of the baboon (Papio sp.): virology; in VAGTBORG, The baboon in medical research, vol. 2, pp. 757–774 (University of Texas, Austin/London 1967).

KAUFMANN, J.H.: Ecology and social behavior of the Coati, Nasua narica, on Barro Colorado Island, Panama. Univ. of Calif. Pub. in Zool., Berkeley/Los Angeles 60: 95–222 (1962).

KEAY, R.W.J.: Vegetation map of Africa (Oxford University, London 1959).

KEMENY, J.J.; SNELL, J.L. and THOMPSON, G.L.: Introduction to finite mathematics (Prentice-Hall, New Jersey 1957).

KENDALL, D.G.: Stochastic processes and population growth. J. roy. statist. Soc. B 11: 230–264 (1949).

KOFORD, C.: Population changes in rhesus monkeys, 1960–1965. Tulane Stud. Zool. 13: 1–7 (1966).

KORTLANDT, A. and KOOIJ, M.: Protohominid behaviour in primates. Symp. zool. Soc. Lond. 10: 61–88 (1963).

KRIEWALDT, F.H. and HENDRICKX, A.G.: Reproductive parameters of the baboon. Lab. anim. Care 18: 361–370 (1968).

KRUUK, H. and TURNER, M.: Comparative notes on predation by lion, leopard, cheetah and wild dog in the Serengeti area, East Africa. Mammalia 31: 1–27 (1967).

KUMMER, H.: Dimensions of a comparative biology of primate groups. Amer. J. phys. Anthropol. 27: 357–366 (1967); Social organization of hamadryas baboons (University of Chicago 1968).

KUMMER, H. and KURT, F.: Social units of a free-living population of hamadryas baboons. Folia primat. 1: 4–19 (1963).

LAMPREY, H.F.: Ecological separation of the large mammal species in the Tarangire Game Reserve, Tanganyika. E. afr. Wildl. J. 1: 63–92 (1963); Notes on the dispersal and germination of some tree seeds through the agency of mammals and birds. E. afr. Wildl. J. 5: 179–180 (1967); Estimation of the large mammal densities, biomass, and energy exchange in the Tarangire Reserve and the Masai steppe in Tanganyika. E. afr. Wildl. J. 20: 1–46 (1964).

LOVERIDGE, A.: Notes on East African mammalia (other than horned ungulates) collected or kept in captivity, 1915–1919. J.E.Afr. Uganda nat.hist.Soc. *16*: 38–42; *17*: 39–69 (1921); Notes on East African mammals. Proc. zool. Soc. Lond. 685–739 (1923).

LUCE, R.D. and RAIFFA, H.: Games and decisions (Wiley, New York/London 1957).

LUMSDEN, W.H.R.: The night-resting habits of monkeys in a small area on the edge of the Semliki Forest, Uganda. J.Anim.Ecol. *20*: 11–30 (1951).

MACDONALD, J.: Almost human. The baboon: wild and tame—in fact and legend (Chilton, Philadelphia 1965).

McNAB, B.K.: Bioenergetics and the determination of home range size. Amer.Nat. *97*: 133–140 (1963).

MAPLES, W.R. and McKERN, T.W.: A preliminary report on classification of the Kenya baboon; in VAGTBORG, The baboon in medical research, vol.2, pp.13–22 (University of Texas, Austin/London 1967).

MARAIS, E.: My Friend, the baboon (Methuen, London 1947).

MAXIM, P.E. and BUETTNER-JANUSCH, J.: A field study of the Kenya baboon. Amer. J.phys.Anthrop. *21*: 165–180 (1963).

MAYNARD SMITH, J.: Group selection and kin selection. Nature, Lond. *201*: 1145–1147 (1964).

MILLER, J.H.: *Papio doguera* (dog-faced baboon) a primate reservoir host of *Schistosoma mansoni* in East Africa. Trans.roy.Soc.trop.Med.Hyg. *54*: 44–46 (1960).

MITCHELL, B.L.; SHENTON, J.B. and UYS, J.C.M.: Predation in large mammals in the Kafue National Park, Northern Rhodesia. Symp.afr.Mam. (Salisbury, Sth Rhodesia 1963).

MOHR, C.O.: Table of equivalent populations of North American small mammals. Amer. midl.Nat. *37*: 223–249 (1947).

MORGAN-DAVIES, A.M.: Progress report on the controlled destruction of olive baboons in the Lake Manyara National Park. Tanganyika Natl.Parks, Arusha (1961 and 1962).

MORGAN, M.T. and TUTTLE, R.H.: Intimate infant–adult male interactions in Rhodesian baboons *(Papio cynocephalus)*. Abstract. Amer.J.phys.Anthrop. *25*: 203 (1966).

MORRIS, D. and MORRIS, R.: Men and apes (McGraw-Hill, New York 1966).

NAPIER BAX, P. and SHELDRICK, D.L.W.: Some preliminary observations on the food of elephants in the Tsavo Royal National Park (East) of Kenya. E.afr.Wildl.J. *1*: 40–53 (1963).

NELSON, G.S.: Schistosome infections as zoonoses in Africa. Trans.roy.Soc.trop.Med. Hyg. *54*: 301–316 (1960).

ODUM, E.P. and KUENZLER, E.J.: Measurement of territory and home range size in birds. Auk *72*: 128–137 (1955).

PATTERSON, B.: The extinct baboon, *Parapapio jonesi*, in the early Pleistocene of northwestern Kenya. Breviora *282*: 1–4 (1968).

PITELKA, F.A.: Numbers, breeding schedule and territoriality in pectoral sandpipers of northern Alaska. Condor *61*: 233–264 (1959).

PITMAN, C.R.S.: A guide to the snakes of Uganda (Uganda Society, Kampala 1938).

ROTH, W.T.: The taxonomy of the baboon and its position in the order of primates; in VAGTBORG, The baboon in medical research, vol.1, pp.3–16 (University of Texas, Austin 1965).

ROWELL, T. E.: The habit of baboons in Uganda. Proc. E. Afr. Acad. *2:* 121–127 (1964); Forest living baboons in Uganda. J. Zool. *149:* 344–364 (1966).

SANDERSON, I. T.: Caribbean treasure (Viking, New York 1939).

SHORTRIDGE, G. C.: The mammals of South West Africa (London 1934).

SMITH, V. S.; FICKEN, R.; LATCHAW, P. and GROOVER, M. E.: The influence of the mature male on the menstrual cycle of the female baboon; in VAGTBORG, The baboon in medical research, vol. 2, pp. 621–624 (University of Texas, Austin/London 1967).

SOUTHWICK, C. H. and SIDDIQI, M. R.: Population changes of rhesus monkeys *Macaca mulatta* in India, 1959 to 1965. Primates *7:* 303–314.

STEVENSON-HAMILTON, J.: Wild life in South Africa (London 1947).

STICKEL, L.: A comparison of certain methods of measuring ranges of small mammals. J. Mammal. *35:* 1–15 (1954).

STOLTZ, L. P. and SAAYMAN, G. S.: Ecology and behavior of baboons in the northern Transvaal (1969, in press).

STRUHSAKER, T. T.: Ecology of vervet monkeys *(Cercopithecus aethiops)* in the Masai-Amboseli Game Reserve, Kenya. Ecol. *48:* 891–904 (1967); Behavior of the vervet monkey *(Cercopithecus aethiops)*; Diss. Berkeley (1968).

SUGIYAMA, Y.: On the division of a natural troop of Japanese monkeys at Takasakiyama. Primates *2:* 109–148 (1960).

TAPPEN, N. C.: Problems of distribution and adaptation of the African monkeys. Curr. Anthrop. *1:* 91–120 (1960).

TAYLOR, C. R.: The minimum water requirements of some East African bovids. Symp. zool. Soc. Lond. *21:* 195–206 (1968).

TOKUDA, K.: A study on the sexual behavior in the Japanese monkey troop. Primates *3:* 1–40 (1961–2).

VAN CITTERS, R. L. *et al.: vide* CITTERS, R. L. *et al.*

VAN WAGENEN, G.: Body weight and length of the newborn laboratory rhesus monkey *(Macaca mulatta)*. Fed. Proc. *13:* 157 (1954).

WAGNER, S. S. and ALTMANN, S. A.: What time do the baboons come down from the trees? An estimation problem. Biometrics (in press [1973]).

WASHBURN, S. L. and DEVORE, I.: Social behavior of baboons and early man. Viking Fund Publ. Anthrop. *31:* 91–104 (1961).

WATSON, A. and MOSS, R.: Spacing as affected by territorial behavior and nutrition in red grouse *(Lagopus l. scoticus)*; in ESSER, Behavior and environment, pp. 92–111 (Plenum, New York 1971).

WILLIAMS, J. G.: Field guide to the National Parks of East Africa (Collins, New York 1968).

WINGFIELD, R.: Some aspects of social behavior of catarrhine primates (especially *Macaca* and *Papio*); Diss. Bristol (1963).

WYNNE-EDWARDS, V. C.: Animal dispersion in relation to social behaviour (Hafner, New York 1962).

ZUCKERMAN, S.: Social life of monkeys and apes (Harcourt-Brace, New York 1932).

4743, 275
46